Marine Diesel Engines

Marine Diesel Engines

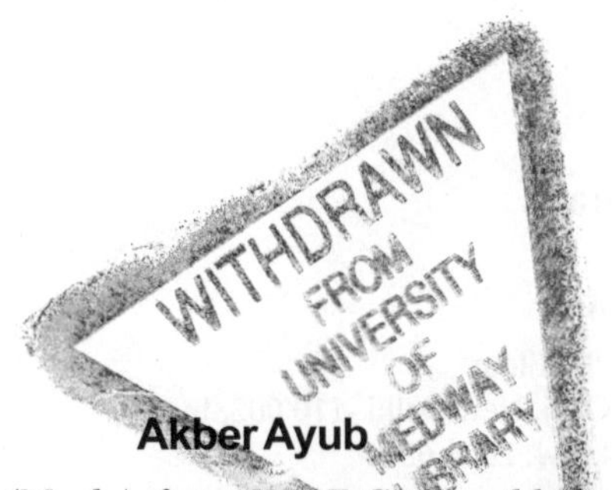

Akber Ayub

B.E.(Mech.) from K.R.E.C. Surathkal
Marine Engg. Competency Certificate MEO-I(A)
Diploma in Software Engineering from Aptech.

CRC is an imprint of the Taylor & Francis Group, an informa business

Ane Books Pvt. Ltd.

Marine Diesel Engines

Akber Ayub

First Published in 2010 by

Ane Books Pvt. Ltd.

4821 Parwana Bhawan, 1st Floor
24 Ansari Road, Darya Ganj, New Delhi -110 002, India
Tel: +91 (011) 2327 6843-44, 2324 6385
Fax: +91 (011) 2327 6863
e-mail: anebooks@vsnl.net
Website: www.anebooks.com

For

CRC Press

Taylor & Francis Group
6000 Broken Sound Parkway, NW, Suite 300
Boca Raton, FL 33487 U.S.A.
Tel : 561 998 2541
Fax : 561 997 7249 or 561 998 2559
Web : www.taylorandfrancis.com

For distribution in rest of the world other than the Indian sub-continent

ISBN : 978-14-3981-414-7

British Library Cataloguing in Publication Data
A catalogue record for this book is available from the British Library

Printed at Thomson Press, Delhi

Dedicated to

My Mother

Mariam Bai Ayub

Preface

Compared to other areas of engineering and technology like traditional streams of engineering and the IT juggernaut, marine engineering is a subject that does not enjoy as yet a mass base in India. Lack of awareness has been a prime reason. In the case of marine engineering, there is a base, but even today, its spread is limited largely to port cities and not too much elsewhere. Realizing the enormous career prospects of a sailing marine engineer and ample opportunities for ex-marine engineers in shipbuilding and utility services including star hotels, a slew of marine engineering colleges have sprung up in various cities in the recent past. Universities have begun opening their doors finally to this field. DG shipping has also stepped in to overview these institutes and provide accreditation. This augurs well for the field of marine engineering and for the shipping industry in general.

However, while Indian marine engineers today hold pride of place in global shipping industry, when it comes to academics, there is a yawning gap. There are very few books, if any, written by Indian authors on marine engineering.

This book is an effort towards filling that gap.

Though introductory in nature, the book contains enough detailed material to be of value to senior students of marine engineering just as much to entry-level marine engineers and trainees. However, sailing marine engineers preparing for their competency examinations will find it most useful. It covers both main propulsion engines and auxiliary engines.

The chapters progress from working principles to construction and design features to operation and maintenance. All engine components and system have been dealt with rather exhaustively. The latest developments in the field, especially in the new electronically controlled engines, have been added towards the end of respective chapters or relevant sections. Hazards inherent in a running engine and the built-in safeguards and fail-safe devices to combat them have been dealt with in a separate chapter. Line drawings and composite diagrams have been used freely to explain concepts and intricacies of design. An added feature is the section on watch keeping, since that is truly a test of the skill and competence of a sailing marine engineer.

With detailed coverage of material, well-defined sections and a clear and focused style of writing, it is hoped the present volume will be of immense help to students and other intended readers.

The author will be glad to receive helpful suggestions and more importantly, intimations of any errors noticed in the book that might have crept in inadvertently.

Akber Ayub

Acknowledgement

I feel indebted to my alma mater KREC, Surathkal, present day National Institute of Technology, for providing the foundation to my technical studies. I am deeply thankful to my aunt Rahima Dawood who tutored me ever so earnestly during my junior college days and who truly nurtured my interest in science.

I wish to acknowledge the help rendered by Bala Sridharan, B.E, F.I.Mar.E, Consultant Marine Engineer, for some of the key inputs and reference material he provided ever so graciously.

Thanks are also due to Rama Swami, B.E, ex-Chief Engineer, for his useful suggestions. I acknowledge the assistance rendered by my publishers Ane Books Pvt. Ltd., and a special thank you to R.K Majumdar, their exceptionally talented graphic artist who worked tirelessly with me on the numerous drawings and diagrams in the book.

Finally, I would like to express my love and gratitude to my wife Sabiha for her patience and understanding all through the long periods while I was working over the manuscript.

Contents

Chapter 1

Types of Ships and Propulsion Machineries

Types of Ships

Merchant ships are generally classified according to the type of cargo they carry. The main types are general cargo ships, ore and bulk carriers, container ships, oil tankers, chemical and gas carriers, reefer ships, Ro-Ro, passenger ships and other miscellaneous vessels.

General Cargo Ship

These vessels, also called freighters, are designed to handle a bulk of international trade. They carry cargo, goods and materials from one port to another. Older ships were fitted with derricks for cargo handling that were replaced with deck cranes in later ships. These vessels can also carry a load of containers or a consignment of bulk cargo.

Bulk Carrier

A bulk carrier or bulker is a merchant ship used to transport unpackaged bulk cargo such as cereals, coal, ore, cement etc. They may carry cargo that is dense, corrosive or abrasive. The distinction between a bulk carrier and a general cargo ship is rather blurred, as today's bulker fleets are built to be multipurpose vessels. A true general cargo ship is therefore rare these days.

Container Ship

These are cargo ships that carry their entire load, mostly manufactured goods, in truck-sized containers. Their design ensures that no space is wasted. The ship capacity is measured in TEUs (Twenty-foot equivalent units), which is the number of 20-foot containers that a vessel can carry, although the majority of containers used today are 40 feet in length. Heavy-duty gantry cranes alongside piers are employed for loading and unloading these vessels. However, smaller ships with capacities upto 2,900 TEUs are often equipped with their own cranes. Large container ships can carry 15,000 containers or more.

Oil Tankers

Oil tankers are ships designed to transport crude oil in bulk. Their capacities range from several hundred tons, employed in coastal runs, to oceangoing supertankers with capacities of several hundred thousand tons. The length of these vessels, termed VLCCs (very large crude carriers), and ULCCs (ultra-large crude carriers) exceed 1500 feet. Many modern tankers feature double hull construction to preclude the danger of leakage in case of damage to the outer hull plates. A complex piping system on the deck, and steam heated stainless steel pipes that run through the cargo tanks, ensure safe loading and unloading of cargo. A wide range of products are carried by tankers. Thus, in addition to oil tankers, there are chemical tankers, gas carriers and even fresh water carriers.

Chemical Tankers

Normally these vessels have a series of separate cargo tanks which are either epoxy coated or painted with specialized zinc paint – to carry safe or less corrosive chemicals. When strong acids such as sulphuric or phosphoric acids are to be shipped, the tanks are made of stainless steel. Chemical tankers generally range from 5,000 to 40,000 dwt (deadweight tons) in size.

Gas Carriers

These are ships designed for transporting natural gases like LPG, LNG and other comparatively innocuous gases like nitrogen, non-toxic refrigerants etc. The gas is generally cooled to about –163° Celsius when it condenses to liquid and greatly reduces in volume. Large spherical tanks made of stainless steel or aluminum are fitted with ‘waffles’ to absorb the thermal contraction when the tanks are cooled down. Alternately, tanks are made of thin primary and secondary membranes of a material called Invar which has almost no thermal contraction. The tanks are located well away from hull plates. Liquid gas in these cryogenic tanks boils at atmospheric pressure and therefore need to be constantly ventilated to avoid build up of pressure. New advances in the field however utilise the 'boil off' gas to fuel the ship's boilers, or re-liquefy it and return it to the cryogenic tanks. Traditionally, steam turbines have been powering gas carriers, in order to make use of the boil-off gas. The capacity of a gas carrier is designated in cubic meters.

Reefer Ships

These are used to transport perishable commodities which require temperature-controlled transportation, mostly fruits and vegetables, meat, sea food, dairy products and the like. Cargo is carried in refrigerated hatches and is generally palletized for easy handling using derricks or cranes. Alternately, a side door on the hull opens out to rest on the quay forming a ramp for trucks to move in which are loaded or unloaded by forklifts. Cargo is also carried in refrigerated containers that are either stacked on the deck or stowed in cell guides inside holds and plugged into the ships mains.

Ro-Ro Vessels

The abbreviation stands for roll-on/roll-off. These are ships designed to carry automobiles or sometimes railway carriages. Ro-Ro vessels have built-in ramps, either at the bow or the stern, which allow the cargo to be conveniently "rolled on" and "rolled off" the vessel when in port. A Ro-Ro's ramp can serve multiple decks. Ro-Ro cargo is typically measured in units of lanes in meters (LIMs), which is calculated by multiplying the cargo length in meters by its width in lanes.

Passenger Ships

Passenger ships include ocean liners, cruise ships and ferries. Ocean liners used to be the traditional form of passenger ships and carried mail and package freight in addition to passengers. They were used for long hauls and so valued speed. More recently however, liners have been mostly replaced by cruise ships and ferries. These have luxurious amenities like swimming pools, theaters, ball rooms, casinos, sports facilities and plush interiors. They are primarily used for pleasure voyages. The size of civilian passenger ships is measured by gross tonnage, which is a measure of its enclosed volume.

Tonnage and Displacement

Tonnage does not refer to the weight of a loaded or empty ship. Rather, it is a measure of the size or cargo capacity of a ship. It is therefore not to be confused with "Ton", which is a unit of weight.

Gross Register Tonnage (GRT) represents the total internal volume of a ship. 1 gross register ton is equal to a volume of 100 cubic feet.

Net Register Tonnage (NRT) represents the volume of the ship available for transporting freight or passengers. Hence, it is the GRT minus the volume of spaces that will not hold cargo such as engine compartment, bridge, crew quarters etc. It was replaced by *net tonnage* in 1994.

Gross Tonnage (GT) refers to the volume of all ship's enclosed spaces, from keel to funnel, measured to the outside of the hull framing. It is therefore larger than GRT.

Displacement is the actual total weight of the vessel and is expressed in metric tons. It is calculated by multiplying the volume of the hull below the waterline (*i.e.*, the volume of water it displaces) by the density of the water. Normal mass density of sea water is 1.025 t/cubic meters. However, different types of water, like summer, fresh, tropical fresh, winter North Atlantic etc have different mass densities.

Deadweight Tons (DWT) is the displacement in tons at any loaded condition minus the lightship weight (weight of the empty ship with no fuel, passengers, cargo, water, etc. on board). It is therefore the difference in weight between a vessel when it is fully loaded and when it is empty, measured by the water it displaces and represents the total weight of fuel, water in tanks, cargo, stores, passengers, baggage, crew and their effects that it carries. The following list gives the dwt of a variety of tankers.

- General purpose tanker: 10,000–24,999 dwt.
- Medium range tanker: 25,000–44,999 dwt.
- LR1 (Large Range 1): 45,000–79,999 dwt.
- LR2 (Large Range 2): 80,000–159,999 dwt.
- VLCC (Very Large Crude Carrier): 160,000–319,999 dwt.
- ULCC (Ultra Large Crude Carrier): 320,000–549,999 dwt.

Marine Diesel Engines

Diesel engines have been powering merchant vessels since the early decades of the 20th century. Today (2009), more than 95 % of merchant ships use diesel engines for propulsion. The diesel internal combustion engine uses the compression ignition system, in which a fine spray of fuel ignites spontaneously when injected into air in the combustion chamber that has been compressed to temperatures higher than its ignition point. The engine operates using the Diesel cycle named after German engineer Rudolf Diesel, who invented it in 1892.

Marine diesel engines employed in ships fall under the following broad classification:

- *By operating cycle:* two-stroke or four-stroke. Most of the slow speed main propulsion engines are two-stroke while the medium-speed and high-speed diesels are four-stroke engines.
- *By construction*: crosshead or trunk. Slow speed engines employ cross-heads while medium speed and high-speed engines are generally of the trunk type.
- *By speed:* Slow speed, medium speed and high speed.

Slow Speed Engines

Any engine with a maximum operating speed up to 300 rpm is termed slow speed. Most large two-stroke slow speed diesel engines however operate at around 120 rpm and form the largest, most powerful engines powering the world's merchant fleet. They are also recognized as the most economical and reliable prime movers. Large two-stroke engines are directly coupled to the propeller.

Medium Speed Engines

Any engine with a maximum operating speed in the range 300-900 rpm comes under the medium speed category. Most of the modern 4-stroke medium speed diesel engines run at around 500 rpm. A gear train coupled to multiple engines reduces the speed to around 120 rpm (most efficient propeller operating speed) to drive the propeller. Alternately, two direct-drive engines power two separate propellers.

High Speed Engines

Any engine with a maximum operating speed above 900 rpm is called a high-speed engine. These engines are used to drive electrical generators on board ships.

Chapter **2**

Diesel Engines

Theoretical Diesel Cycle

The thermodynamic changes occurring inside the cylinder of a compression ignition diesel engine may be best expressed by studying its theoretical heat cycle. The term 'cycle' here refers to one complete sequence of events required to produce one power stroke in an engine; stroke being the full distance the piston moves between each end of its travel. Compression of the air inside the cylinder increases its temperature to a point where ignition occurs spontaneously when fuel is injected (most diesel engines have a compression ratio of 16:1 to 23:1). Fuel injection and combustion are controlled to give constant pressure combustion as the expanding gases keeps pace with the increase in volume caused by piston travel. Thus, combustion is said to occur at constant pressure. This is followed by constant-volume expulsion of the gases.

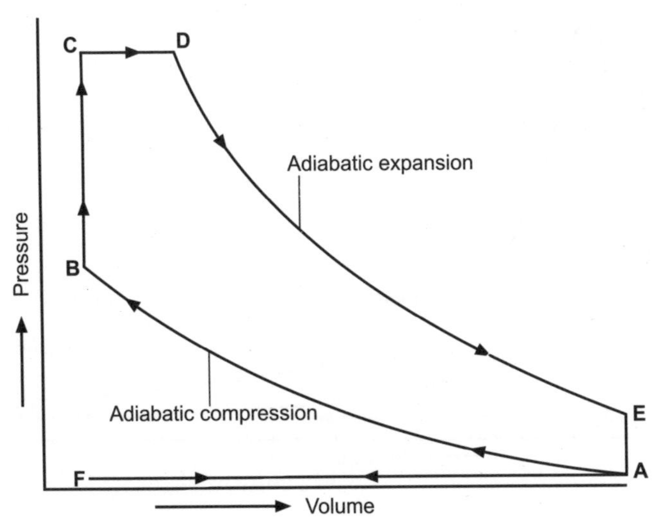

Fig. 2.1 Theoretical heat cycle PV graph.

Referring to Fig. 2.1, air admitted to the cylinder under natural aspiration is compressed adiabatically (without loss or gain of heat). Beginning of compression is denoted by point A and end by point B. Fuel injection begins at point B and the resulting ignition and combustion adds heat to the cycle partly at constant volume, because of near instantaneous combustion – denoted by line BC – and partly at constant pressure – denoted by line CD. Expansion begins at point D and continues, again adiabatically, till point E, when heat in the exhaust gases is expelled at constant volume, represented by line EA. In the case of a four-stroke engine, once the exhaust gas pressure drops to near atmospheric, further movement of the piston during the exhaust stroke expels the reminder of the gases at atmospheric pressure. During the next suction stroke air is once again drawn into the cylinder at atmospheric pressure. These two processes are represented by the horizontal line AF.

This is a **hypothetical indicator diagram** representing what is called an 'air standard cycle' that assumes constant specific heats; frictionless adiabatic, *i.e.*, isentropic expansion and compression; charge air mass remaining unaffected by the injected fuel etc. The efficiency of this theoretical cycle would be around 55 to 60 %. Heat lost in the exhaust gases and conducted through the piston, cylinder and cylinder head account for the rest 40 to 45 % – the lost energy that cannot be recovered. An actual indicator diagram however, will be different from this ideal state.

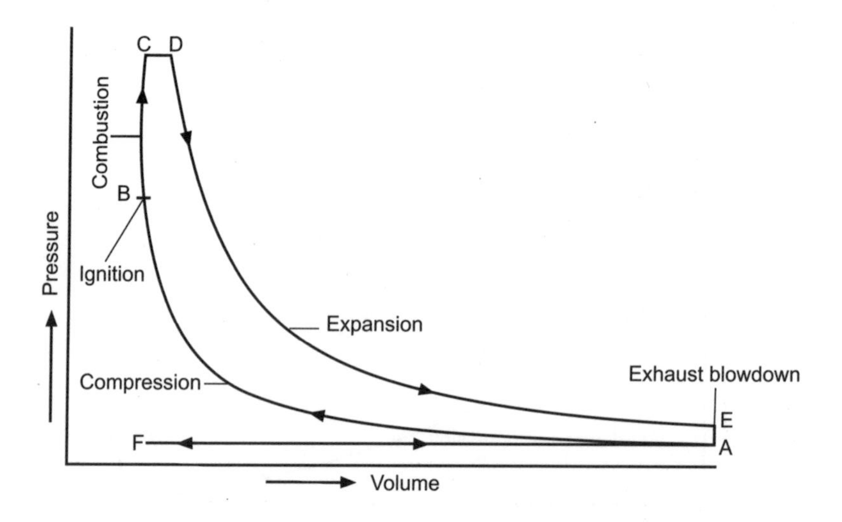

Fig. 2.2. (a) Attainable perfect indicator diagram.

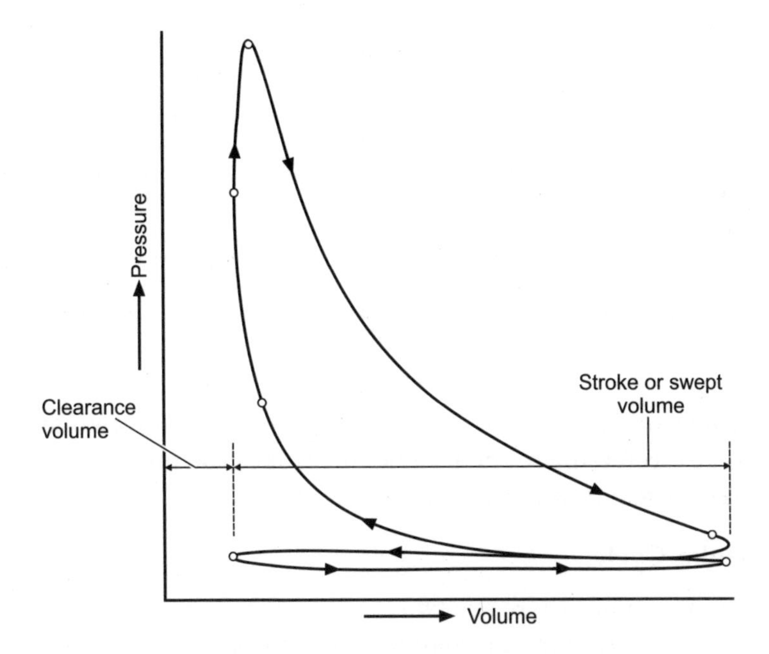

Fig. 2.2 (b) Working indicator diagram.

Since both the compression and expansion processes would be non-adiabatic due to dissociation of heat and the various heat losses – thus resulting in lower attainable terminal pressure and temperature – the compression and expansion curves are lower compared to the theoretical diagram. Again, there is the further mechanical losses due to friction in the various engine mechanisms and the power expended for essential drives like camshaft, pumps etc. Fig. 2.2 (a) represents an **attainable diagram** of an actual diesel engine – assuming the engine is in perfect condition in all respects. This represents what is attainable in an ideal case. However, since both the inlet and exhaust valve operations are non-instantaneous, the corresponding corners in the diagram will be typically rounded. The actual expansion is also non-adiabatic with other deviations from the ideal case like, some combustion taking place even in the expansion stroke, variable specific heats and mixing of air and gas etc. Therefore a working diagram will be as shown in Fig. 2.2 (b). Consequently, the indicated thermal efficiency of an engine – denoting the percentage of heat energy, obtained by burning fuel in the cylinder, which is converted in to useful work – is generally between 40 to 45 %, while the brake thermal efficiency – denoting the usable power available at the end of the crank shaft – would be between 30 to 40 %. A simple heat balance of a diesel engine is shown in Fig. 2.3, representing the input, output and the various losses.

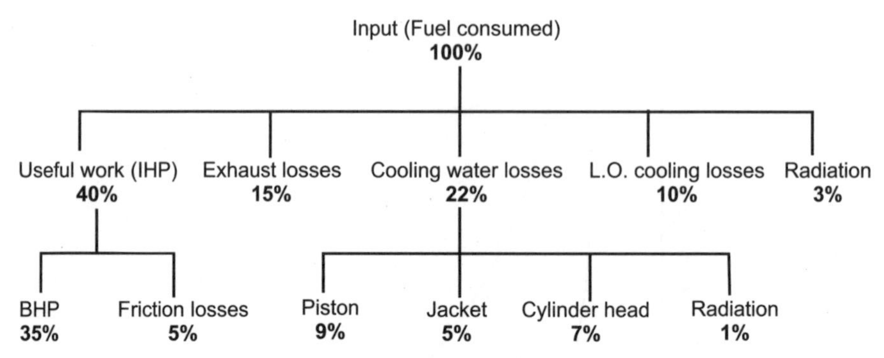

Fig. 2.3 Simple heat balance

The operating cycle of an engine can be either two-stroke or four-stroke – referring to the number of strokes the piston completes in each engine cycle to generate one power stroke.

Two-stroke Cycle

As the name implies there are two strokes of the piston for each cycle of the engine, resulting in one complete revolution of the crankshaft. Air intake, compression and ignition occur in one stroke, while combustion, expansion and exhaust take place in the next. Thus, two consecutive strokes of the piston – that may be termed as *compression stroke* and *expansion or power stroke* – complete one revolution of the engine. The various operations like air intake, compression, ignition etc repeat in the same sequential order for each revolution. They occur when the piston reaches a corresponding position during its stroke and can be represented by circles in a timing diagram – in terms of angles of crank position measured from the top dead centre (TDC) or the bottom dead centre (BDC). Timings differ to some extent for different engines depending on variables like stroke to bore ratio, connecting rod length to crank length ratio, engine power and speed.

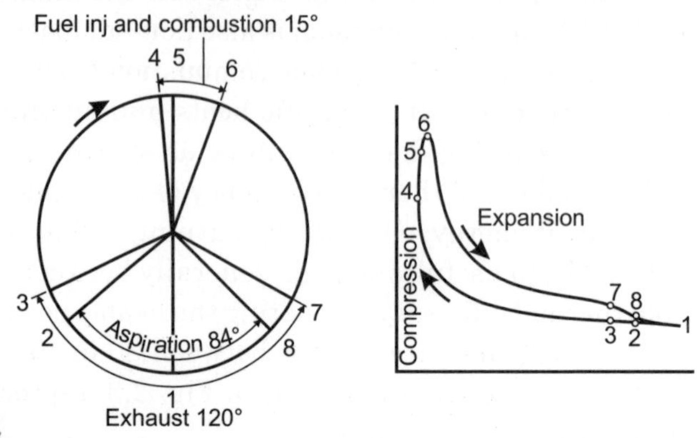

Fig. 2.4 (a) Timing diagram (naturally aspirated) and two-stroke cycle.

Fig. 2.4(a) shows the timing and indicator diagrams for a two-stroke loop-scavenged turbocharged engine. For proper operation of a two stroke engine, air pressure in the scavenge air manifold, provided by an exhaust gas turbocharger, is used to 'scavenge' each cylinder as a series of ports is exposed in the cylinder walls that are connected to the manifold. This scavenge air expels the spent gases from the cylinder through the open exhaust ports and loads fresh charge air for the next stroke (sector 1-2), in what is called the loop-scavenging system (as opposed to the uni-flow poppet valve controlled system, where gases are expelled through a poppet valve mounted on the cylinder head and operated either by hydraulic pressure from a cam driven actuator or a cam driven push rod). Thus, scavange ports are open from position 8 till position 2. This completes aspiration or air induction. Next, the upward movement of the piston blocks off the cylinder-side scavenge ports. Some air within the cylinder will leak out along with the last of the gases through the still open exhaust ports (in some engines both the exhaust and the scavenge ports are at the same height to eliminate this charge air loss). The advancing piston next closes the exhaust ports trapping the charge air in the cylinder (sector 2-3) and compression begins. At or near the top of the compression stroke (sector 3-4), fuel is injected into the cylinder from the fuel injector and near-instantaneous combustion results as the fuel ignites and combustion proceeds with accelerated intensity (sector 4-5-6). The piston is then forced down by the rapidly rising pressure as combustion proceeds and continues for some distance even after the injection stops (position 6). The downward moving piston next uncovers the exhaust ports (position 7). The spent charge is then expelled and as the piston continues downward due to inertia, the scavenge ports are once again exposed (position 8) and the cycle repeats. Position 1 represents the BDC and position 2 the TDC. The angle of the crank at which each event occurs is represented by numbers and the period of operation in degrees. It will be seen that opening and closing of the ports are symmetrical about the BDC since these events are controlled by the piston. The exhaust opening is fixed at the most effective position that would minimise the loss of charge air through the exhaust port.

Fig. 2.4 (b) shows the timing diagram for a four-stroke poppet valve type turbocharged engine. Since the exhaust valve can be controlled independent of the piston, the exhaust begins significantly earlier in this system. This allows the pulse energy of the exhaust gas to be effectively utilised in the turbocharger. Additionally, since the closing position of the valve can also be controlled, this is fixed at the optimum position to minimise the loss of charge air into the exhaust. These two changes together with the pressure charging allow efficient scavenging. The large mass of air flowing into the cylinder also helps in cooling the internal components.

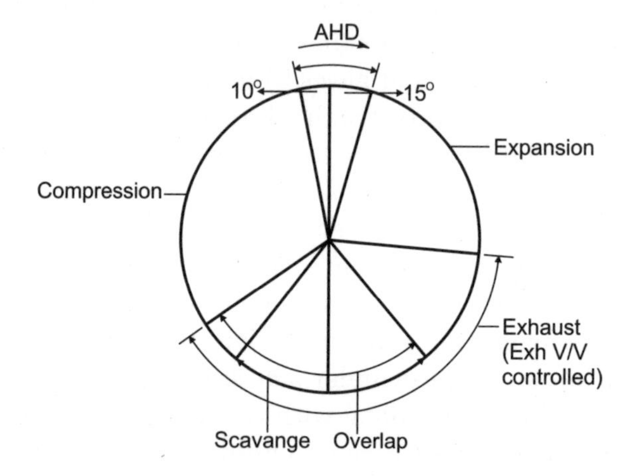

Fig. 2.4 (b) Timing diagram (turbo-charged engine)of a four-stroke engine.

Two-stroke Engine

Fig. 2.5 (a) shows the cross-sectional view of a Sulzer main diesel engine. The cylinder head seals the top of the cylinder and carries the starting air valve, fuel injector, a relief valve and indicator valve.

Starting air valve feeds compressed air in to the cylinder to push the piston down for cranking the engine while starting. High-pressure fuel is delivered to the fuel injector by an individual plunger in the fuel pump that is driven by the camshaft (or, in latest engines, through a solenoid valve from a common rail, carrying fuel oil under high pressure). A train of spur gears, or a chain drive, transmits the rotary motion from the crankshaft to the camshaft. The relief valve releases dangerously high pressure from the cylinder in case of an irregular combustion due to a leaking injector or a mistimed injection. For taking engine power and peak-pressure diagrams an indicator is mounted on the indicator valve.

The cylinder liner, made of alloy cast iron, sits inside a cooling water jacket. This enables efficient cooling and allows easy replacement of a worn liner. The annular space between the liner and the engine block – formed by cast iron jackets bolted together – is sealed by silicone O-rings that withstand high temperatures. Scavenge and exhaust ports in the bottom half of the liner facilitate air intake and expulsion of exhaust gases.

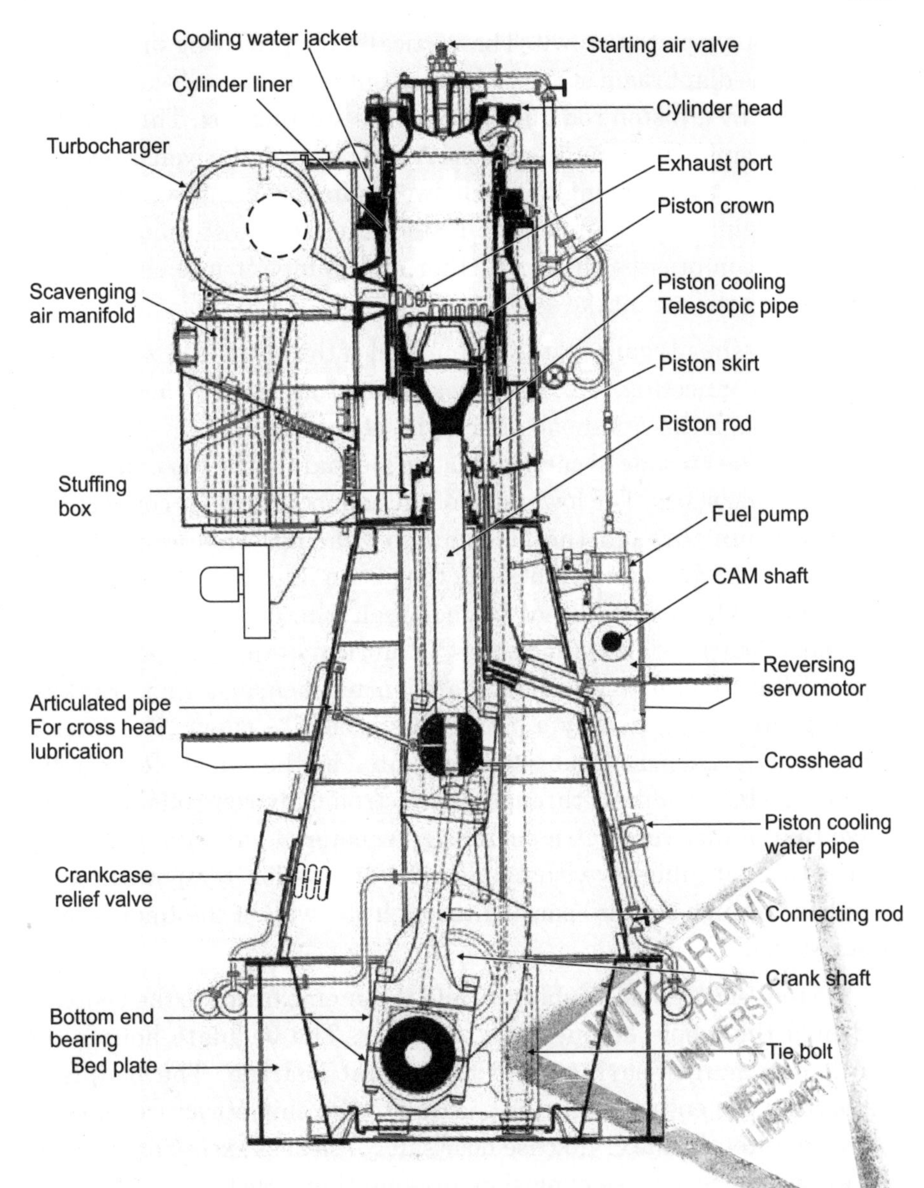

Fig. 2.5 (a) Sulzer RND two-stroke engine section.

The two-piece piston (alloy steel piston crown and cast iron skirt bolted together) has number of annular chrome-plated grooves on the outside that carry piston rings ensuring an efficient seal between the high pressure combustion gases and the under-piston scavenging spaces. The long piston skirt controls the exhaust timing by uncovering the exhaust ports in the liner. Exhaust ports are placed slightly higher than the scavenge ports facilitating efficient loop scavenging. The piston is cooled by circulating water (or oil) that is led through telescopic tubes into the hollow piston crown. The vertically reciprocating piston rod is sealed at the diaphragm at the base of the entablature (enclosed scavenge air spaces) by a piston rod gland called the stuffing box. This prevents air from escaping into the crankcase. A turbocharger (driven by exhaust gases, and in turn driving an air blower) supplies compressed air into the scavenging spaces under the piston. The downward movement of the piston compresses the air further and pumps it into the cylinder liner during the air intake stroke.

The crosshead connects the bottom end of the piston rod with the top end of the connecting rod. The threaded bottom end of the piston rod passes through the crosshead pin and is secured against it with a single large nut. The crosshead carries the two journals of the top end bearing and the guide shoes. The lower end of the connecting rod is connected to the crank pin through the bottom end bearing. This arrangement converts the reciprocating motion of the piston into rotary motion of the crankshaft which is made of forged-steel. The crosshead, including crosshead bearings and guide shoes, is lubricated through articulating pipes. White metal lined crankshaft journal bearings and crankpin bearings are lubricated by a circulating lube oil system. The liner is lubricated by a separate lubricator actuated by the camshaft, or in the case of modern engines, through an electronically controlled solenoid valve that allows volumetric metering. A metered quantity of cylinder lubricating oil (able to withstand high combustion temperatures) is supplied through "quills" passing through the wall of the liner into its inner surface.

Tie rods secure the entablature and the engine frame to the bedplate and hold the whole engine block together. The bedplate houses the crankshaft main bearings and the thrust bearing. The bedplate, supporting the entire engine, is secured to the ship structure. A relief valve mounted on the crankcase door safely releases excess pressure in the event of a crankcase explosion, pre-empting structural damages to the engine.

A B&W KEF engine that uses a push rod operated exhaust valve is shown in Fig. 2.5 (b). In this earlier model, the cylinder liner carried only the scavenge ports at its lower end and a push-rod operated exhaust valve and uniflow scavenging (unidirectional flow of gases pushed up by the piston and expelled through the exhaust valve in the cylinder head) ensured better scavenging efficiency with the least mixing of air and gas. The pulse exhaust system utilised the pulse or impulse energy of the pressure wave formed by the expulsion of exhaust gases from each cylinder to drive the turbocharger connected to three cylinders. Bedplates were generally fabricated and crankshafts were semi-built up. For reversing, the whole camshaft was rotated relative to the fuel pumps and exhaust valve cams in order to re-time them. Fuel injection cooling was employed since the engines burned heavy oil.

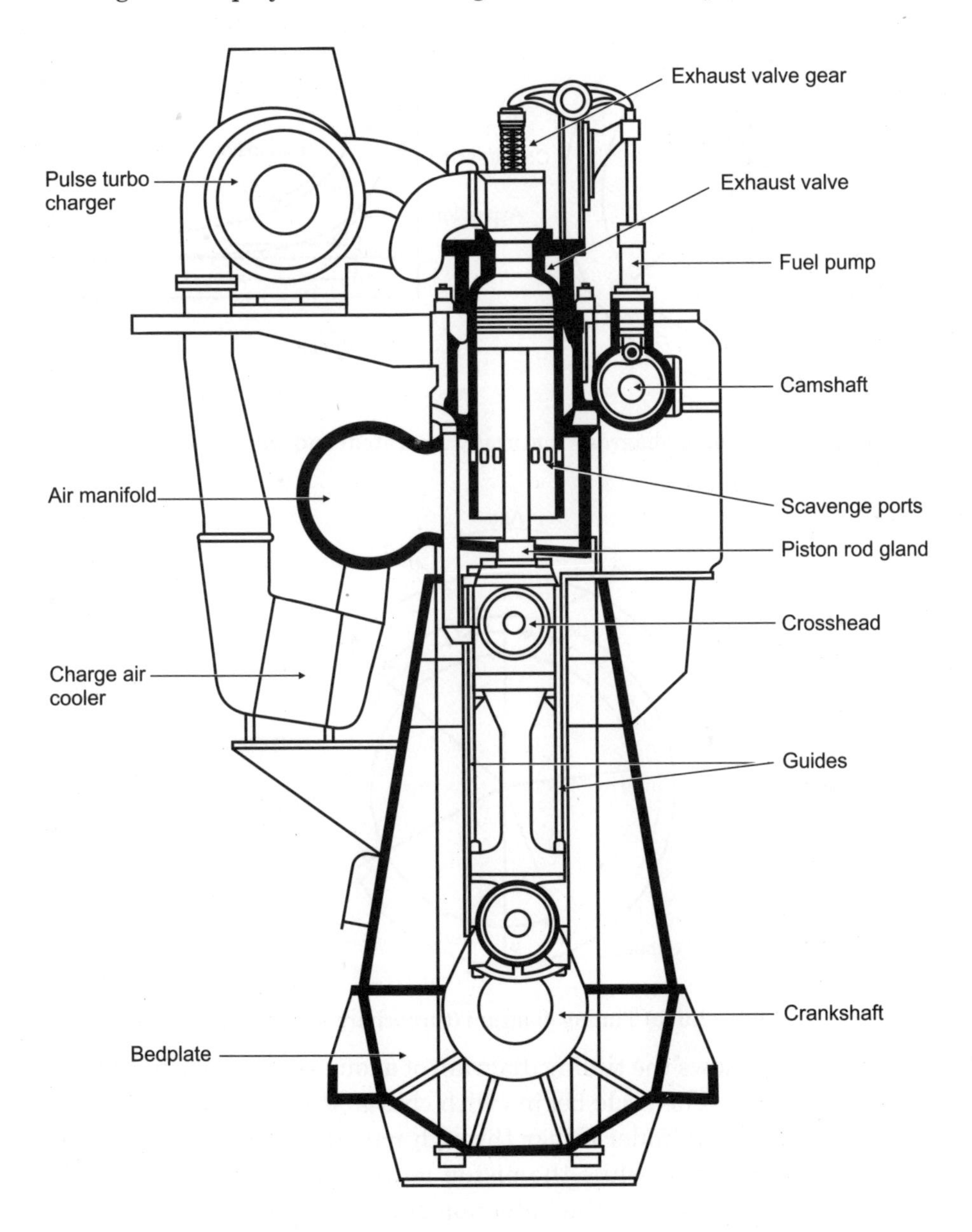

Fig. 2.5 (b). B & W KEF engine section.

Four-stroke Cycle

The sequence of operations in a four-stroke cycle are induction, compression, expansion and exhaust, accounting for four strokes of the piston and resulting in two revolutions of the engine.

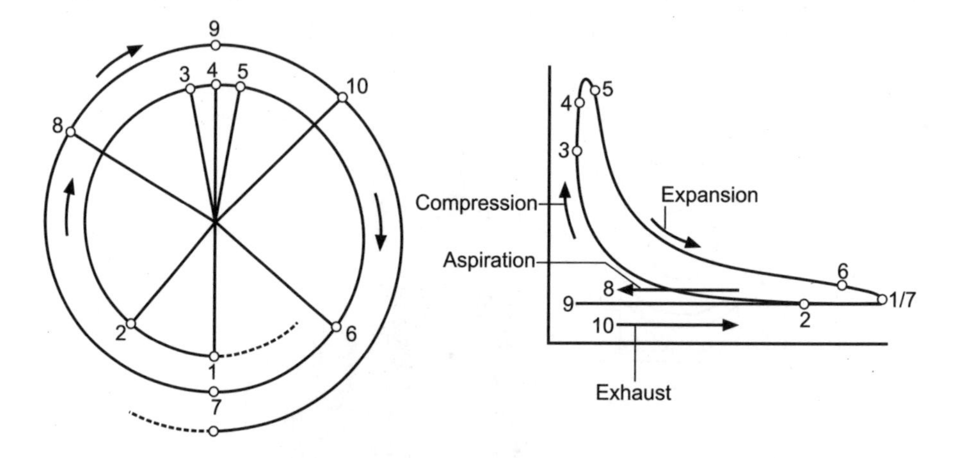

Fig. 2.6 (a) Timing diagram (naturally aspirated) and four-stroke cycle.

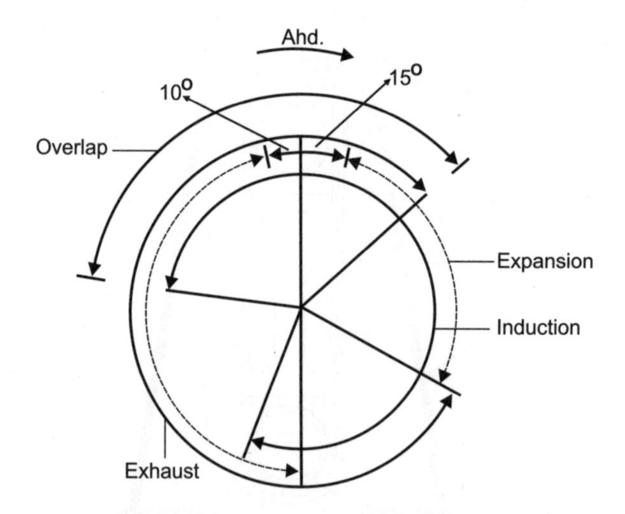

Fig. 2.6 (b) Timing diagram (turbocharged engine).

Fig. 2.6 (a) shows the timing diagram of a four-stroke cycle naturally aspirated engine. The cycle begins with charge air intake, where air is admitted into the cylinder either through natural aspiration or by the use of a turbocharger, while the piston moves downwards and the air inlet valve is open. This is the induction or the suction stroke (sector 1-2). The valve remains open while the piston crosses the BDC and starts moving up. Around 30 degrees after BDC, the inlet valve closes (position 2) trapping the charge air inside the cylinder. Next, the upward moving piston (pushed by the momentum of the crankshaft and flywheel) compresses the air to a fraction of its original volume by the end of the compression stroke (sector 2-3). Fuel is injected into the cylinder before the piston reaches TDC (to ensure finite time for the fuel to ignite called the ignition lag or delay) through the fuel injector. The fuel ignites, setting off combustion and the rapidly increasing pressure drives the piston downwards, after it has crossed the TDC (injection stops around 35 degrees after TDC, depending on the engine's load). This forms the expansion or power stroke (sector 5-6). Before the piston reaches the BDC the exhaust valve opens (position 6) and the spent gases begin to be forced out of the cylinder. During the next upward stroke, the piston expels the rest of the exhaust gases, in what is called the exhaust stroke, till the exhaust valve closes (position 8). Subsequently, the air inlet valve opens while the exhaust also remains open creating an overlap that ensures efficient scavenging. The exhaust valve closes some 54 degrees after the TDC and air is drawn into the cylinder during the downward stroke to begin aspiration for the next cycle.

Fig. 2.6 (b) shows the crank timing diagram of a four-stroke turbocharged engine. Compared to the timing diagram of the naturally aspirated engine it can be seen that this system has a large degree of valve overlap which, together with turbo charging allows very efficient scavenging, the large air mass flow also helping in cooling the internal components.

Four-stroke Engine

Fig. 2.7 shows the cross section of a four-stroke trunk-piston engine. The name "trunk piston" refers to the piston skirt or trunk, its purpose being the same as that of a cross head in a two-stroke engine. It transmits the transverse thrust caused by connecting-rod angularity to the side of the cylinder liner, in the same way the crosshead slipper transmits the thrust to the crosshead guide. In a four-stroke engine there is no diaphragm separating the crankcase and the liner and piston. Separate lubricating oil for the liner is therefore not possible in four-stroke trunk-piston engines.

The cylinder head covering the top of the liner houses the inlet and exhaust valves in addition to the fuel injector, air-starting valve and indicator cock. The cylinder liner therefore does not have the inlet and exhaust ports machined into it. Charge air is led into the cylinder by the inlet valve through passages in the cylinder head and the exhaust gases are expelled from the cylinder by the exhaust valve, again through cylinder head passages. This is the primary difference in the construction of the four and two strokes engines.

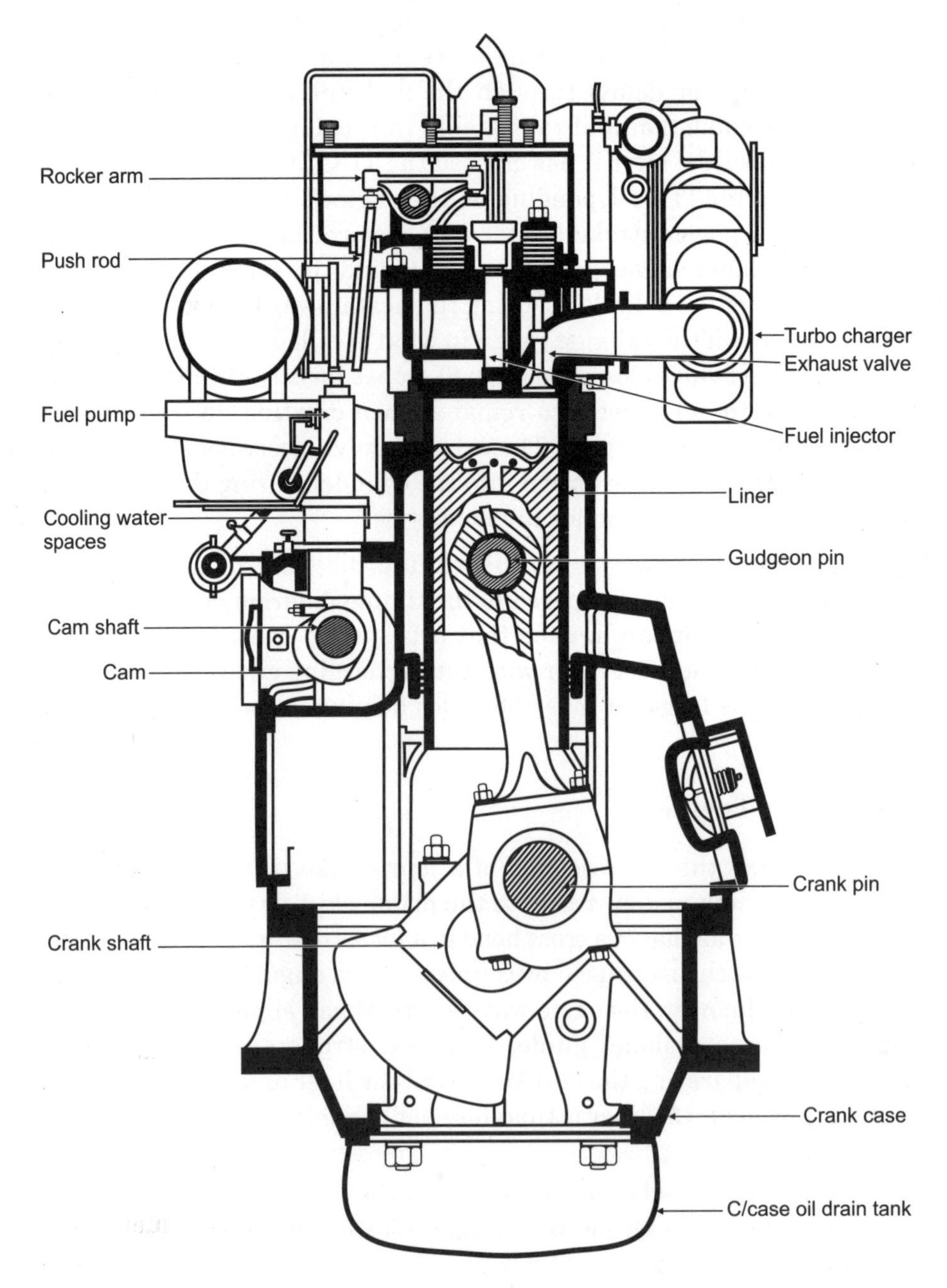

Fig. 2.7 Cross-sectional view of a four-stroke B & W Engine.

The piston is connected to the top end of the connecting rod by a gudgeon pin, and the connecting rod bottom end is connected to the crank pin by the big end bearing. These linkages convert the reciprocating motion of the piston into rotary motion of the crankshaft. A train of spur gears (or a roller chain drive) transmit the drive from the crankshaft to rotate the camshaft which carries separate cams to operate the inlet and exhaust valves, in addition to the fuel pump and air starting valves. Push rods connected to rollers (cam followers) sitting on the cams operate rocker arms which in turn actuate the inlet and exhaust valves mounted on the cylinder head – timed to operate at precise points in the cycle. The valves are normally kept shut by helical steel springs and are pushed down to open, by the opposite end of the pivoted rocker arm. The crankshaft is housed inside the crankcase that sits on the bedplate. The bedplate also carries the crankshaft bearings. Cylinder and cylinder head are cooled by water circulating through cooling passages running inside the cylinder head and around the liner.

The rotational energy of the crankshaft imparts momentum to the flywheel which not only provides the brake horse power for end use, but also accomplishes compression of the charge air and overcomes mechanical losses incurred in the cycle like driving the camshaft for operation of inlet and exhausts valves, fuel injector pump etc.

Comparison of Two-Stroke and Four-Stroke Engines

Theoretically, a two-stroke engine with one power stroke for every revolution should develop twice the power of a four-stroke engine which produces one power stroke for every two revolutions of the engine – for the same power rating. However, poor scavenging efficiency of a low-speed two-stroke engine and other losses reduce the power advantage considerably. Still, the two-stroke diesel engine burning low-cost fuels offers the highest thermal efficiency and is generally recognized as the most economical and reliable prime mover. Structurally too, due to the absence of valve operating gear, a two-stroke engine is considerably lighter, offering a better power-to-weight ratio. These engines are also usually narrow and tall due to the crosshead. Large two-stroke engines used for main propulsion of ships are capable of producing over 100,000 bhp and operate in the range of 90 to 120 rpm. They can be up to 15 m tall, and can weigh over 2000 tons. They typically run on cheap low-grade "heavy fuel", also known as "bunker" fuel, which requires heating during storage and pumping due its high viscosity. Diesel fuel is used for starting the engine and also while maneuvering the ship. Latest engines using the common rail system of fuel injection though use only heavy fuel.

Medium speed four-stroke engines are widely used to power smaller ships and ferries, but since they normally run at 400 to 900 rpm, require a flexible coupling and reduction gearing to drive the propeller. Either two or four engines are sometimes coupled together to a gearbox to drive single or twin propellers. This offers increased reliability and easier maintenance apart from the possibility of shutting down some engines when the ship runs at partial load. Furthermore, variable pitch propellers obviate the need for reversible engines.

Four-stroke engines are used almost always for electric power generation on vessels–coupled to generators and run on diesel fuel alone (light distillate marine diesel oil). Medium and high speed engines may be either in-line or have a V configuration. V-engines have two banks of cylinders at an angle powering a single crankshaft housed inside a common crankcase. These can be found in many V configurations: V8, V12, V16 and V20. Compared to large direct drive two-stroke engines, an indirect drive four-stroke engine is smaller in size and therefore even with a train of engines, the machinery spaces could be considerably smaller for the same power output. However, recent developments have produced bigger four stroke engines with the ability to burn heavy oil. This has made the four stroke engine too an attractive option as a main propulsion engine.

Indicator Diagrams

Indicator diagram is a graph of the pressure inside the cylinder plotted against the changing volume of the cylinder. Indicator diagrams are generated using the engine indicator attached to the indicator cock on the cylinder head. The power developed within the engine cylinder – called the *indicated power* – can be calculated from the area of the traced diagram, since it is directly proportional to the work done during the cycle. This is higher than the *brake power* or the *shaft power* which is available at the crankshaft as the engine output – measured using a torsion meter aboard ship or brake drum on the shipyard test bed.

There is no marked difference between the diagrams for two-stroke and four-stroke engines. Four types of indicator diagrams can be obtained for both. **Power card** is taken with the indicator drum rotating in phase with piston movement. This is used not only to calculate the power produced in a cylinder but also its mean indicted pressure (mip). The highest point in the diagram indicates the maximum pressure. A **draw card** is taken the same way as a power card but with the indicator drum 90° out of phase with the piston stroke. This readily indicates pressure

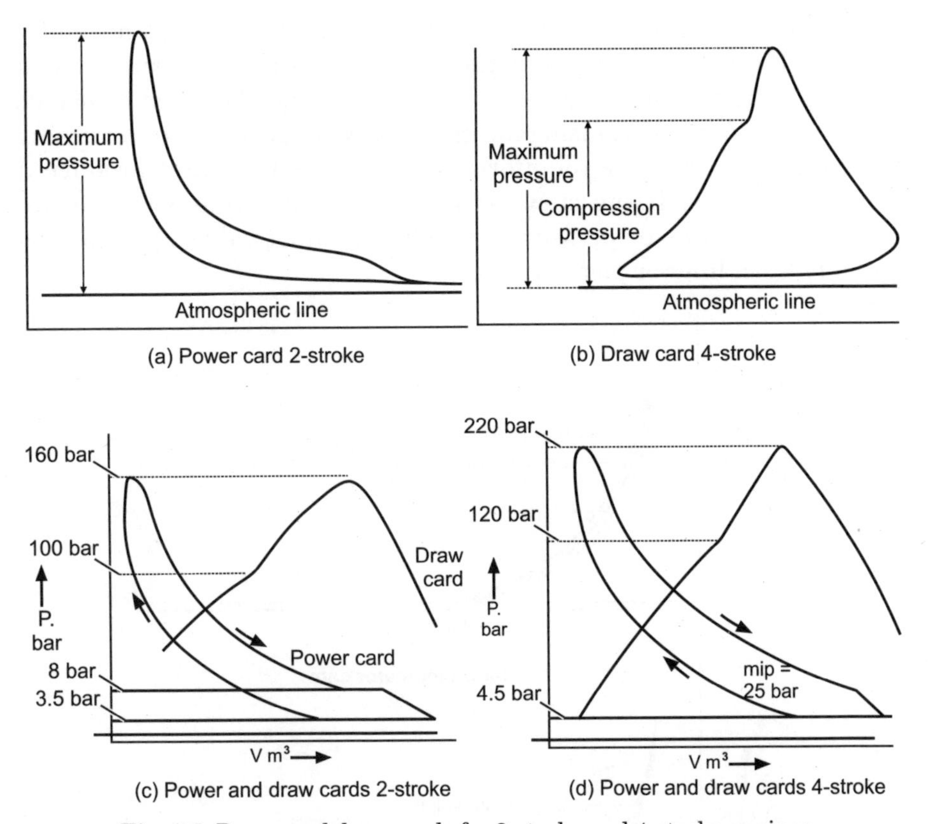

Fig. 2.8 Power and draw cards for 2-stroke and 4-stroke engines.

variations during the combustion or expansion stroke making it easier to identify faulty injection timing or a leaky or choked injector. Ignition delay (from 0.5 to 10 milliseconds) between start of injection and combustion pressure rise can be clearly seen in draw cards. Shorter the delay higher is the ignition quality of the fuel – expressed either by the *diesel index number* or the *cetane number*. Fig. 2.8 (a) and (b) show power card for two stroke engines and draw card for four stroke engines respectively. For the four-stroke cycle, the suction and exhaust strokes are shown by a horizontal line, since pressure changes during these strokes are too low to be recorded and, as can be seen, they do not have any effect upon the cycle. Figs. (c) and (d) show power and draw cards superimposed, for two and four stroke engines respectively. Indicative pressures are shown on the sketches that are nearly similar to each other except that the points indicating start of compression and peak pressure in a four stroke engine are clearly indented as compared to that in the two stroke engine, which are smoother. **Compression card** is taken the same way as a power card but with the fuel shut off for the cylinder (by opening the injector priming valve in a traditional engine). The height of the diagram indicates the maximum compression pressure.

Curves representing compression and expansion should normally coincide. If they do not, it would indicate number of irregularities, like a time-lag in the indicator drive or phase difference between camshaft and crankshaft. Synchronizing the indicator with the engine or correcting the cam shift on the circumference of the shaft should produce a coinciding curve. If the curve is lower in height that would mean insufficient charge air pressure or a leaky exhaust valve, apart from a possibly worn liner or piston rings.

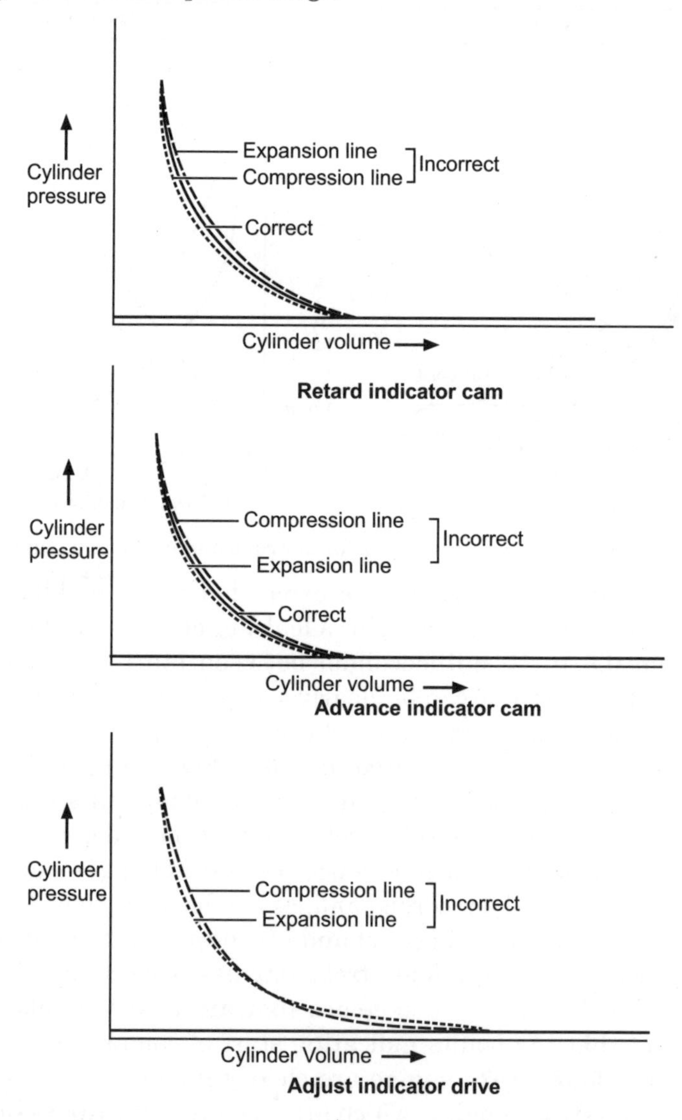

Fig. 2.9 Compression diagram.

Light spring diagram is also same as a power card but taken with a lighter indicator spring. This readily shows pressure changes during scavenging and exhaust, making it easy to detect faults in these two operations – arising from piston blow past and early or late injection.

Compression and firing pressures vary according to engine type and these would be given in the engine manual. It is important that these are not exceeded, especially when the ship is running with a full load. Indicator diagrams help in maintaining correct pressures.

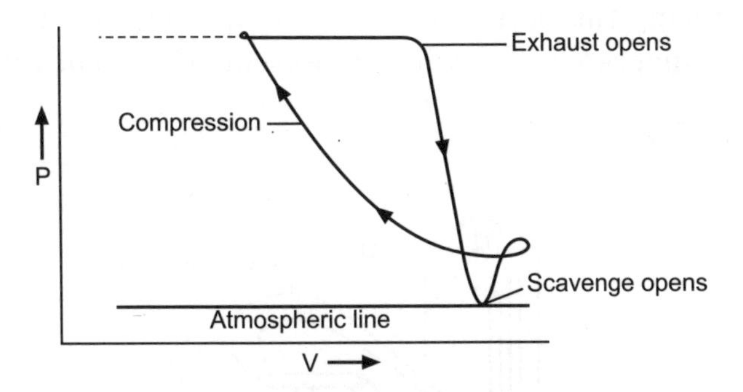

Fig. 2.9.1 Light spring diagrams.

Power Calculations

$$\text{Indicated horse power (ihp)} = \frac{N \times p \times l \times a \times n}{4500}$$

Where,

N = no. of cylinders;

p = mean indicated pressure in kgf/sq.cm;

l = engine stroke, in metres;

a = cross-sectional area of one cylinder, in sq.cm;

n = working strokes per minute (n = rpm/2 for 4-stroke and n = rpm for 2-stroke);

4500 = constant for conversion into metric, in kg metres per min.

On a ship, since all cylinders of the engine would be identical, cylinder dimensions can be replaced by a constant.

Hence, indicated horsepower (ihp) = mip × R × C (per cylinder).

Where,

mip = mean indicated pressure in kgf/sq.cm

R = engine RPM

C = cylinder constant

ihp of engine = ihp per cylinder × no. of cylinders.

mip is obtained by measuring the area of the draw card in sq. cm using a planimeter (an instrument for measuring the area of a plane figure by tracing the perimeter of the figure) and dividing it by the length of the diagram in cm, which represents the swept volume.

Engine Indicator

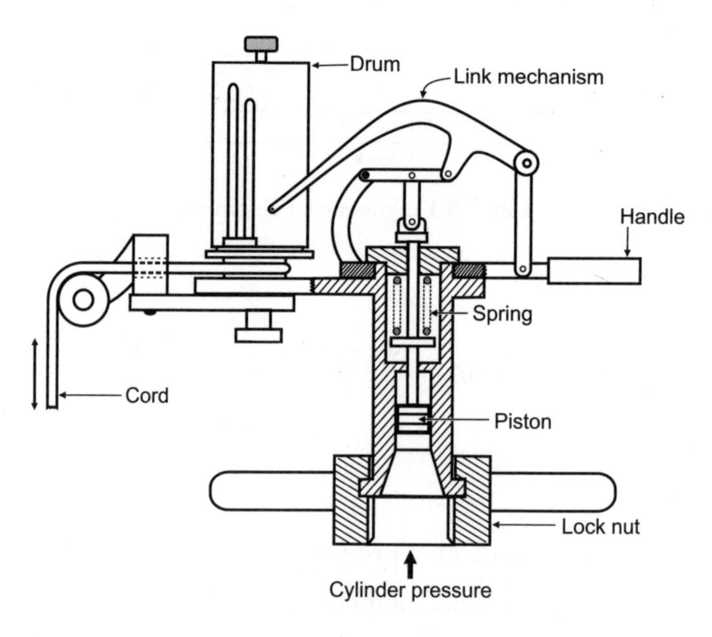

Fig. 2.9.2 Engine Indicator.

The device consists of a spring-loaded piston exposed to the engine cylinder pressure. Attached to the piston is a pen through a parallel link mechanism. The pen touches a sensitised paper wound around a drum. A cord around the drum attached to an engine stroke synchronising mechanism rotates the drum to and fro. The reciprocating mechanism consists of an eccentric mounted on the camshaft that operates a spring-loaded push rod to which the chord is attached.

The indicator valve of a cylinder is first blown through to clear any deposits. Next, the indicator is screwed on and the valve opened. When the pen is held against the paper it traces a diagram for one cycle of the cylinder. If the valve is now shut off, rendering the pen stationary, the pen traces a horizontal line representing the swept volume. If the drum movement is stopped by disengaging the chord and the valve opened once again a vertical line will be traced on the paper representing peak pressure.

Though this type of mechanical indicator has been in use for a long time, it however suffers from some inaccuracies due to friction and inertia of its working gear, and has been found to be increasingly inadequate to monitor the performance of today's high-powered engines.

Electronic indicator is a new development that has largely overcome these shortcomings. Cylinder pressure is measured by a piezo-electric transducer attached to the indicator cock (a transducer being a device that produces electrical, hydraulic or pneumatic signals from a different kind of measured signal like pressure, mechanical stress, magnetic flux etc). The corresponding piston position is picked up by a magnetic sensor fixed close to the flywheel – part of a crank-angle trigger system that determines the TDC for each cylinder and automatically identifies which cylinder is being measured. These data collected for number of cycles are fed into a microprocessor which calculates the mean indicated pressure and the engine power. Additionally, with the help of a pressure transducer installed in the injector fuel line, it is possible to get an accurate reading of the injection timing and pressure. Besides, engine load diagrams can also be produced. Typically, values are obtained for the compression pressure, maximum pressure, mean indicated pressure and scavenge air pressure, as well as the engine speed and engine output for each cylinder and the engine as a whole. These performance data can be viewed instantly on a PC and printed, allowing continuous monitoring of engine performance and trouble shooting. Stored in the database, they can be retrieved subsequently for calculation and display.

Typical Marine Engines

Two-stroke

Sulzer RTA series engines are large slow-speed, two-stroke, crosshead type, long stroke engines used as the main engine for propulsion of fast, large container ships. Cylinder bores of 840 mm, strokes of 2400 mm and operating speeds of 100 rpm are typical. They employ uni-flow scavenge system and constant pressure turbo charging. Exhaust valve is positioned centrally on the cylinder head inside a cage and is rotated as the exhaust gases flow past vanes fitted to its spindle. It is opened by hydraulic pressure from a cam-driven actuator. This series of engines is an advanced version of the earlier RND series. These earlier engines employed loop scavenging through exhaust ports in the liner skirt, which meant close proximity of both inlet and exhaust ports on the liner – and the consequent sharp temperature gradient. Constant pressure turbo charging augmented by under-piston charge air compression was used. They had cam-driven, valve-timed fuel pumps.

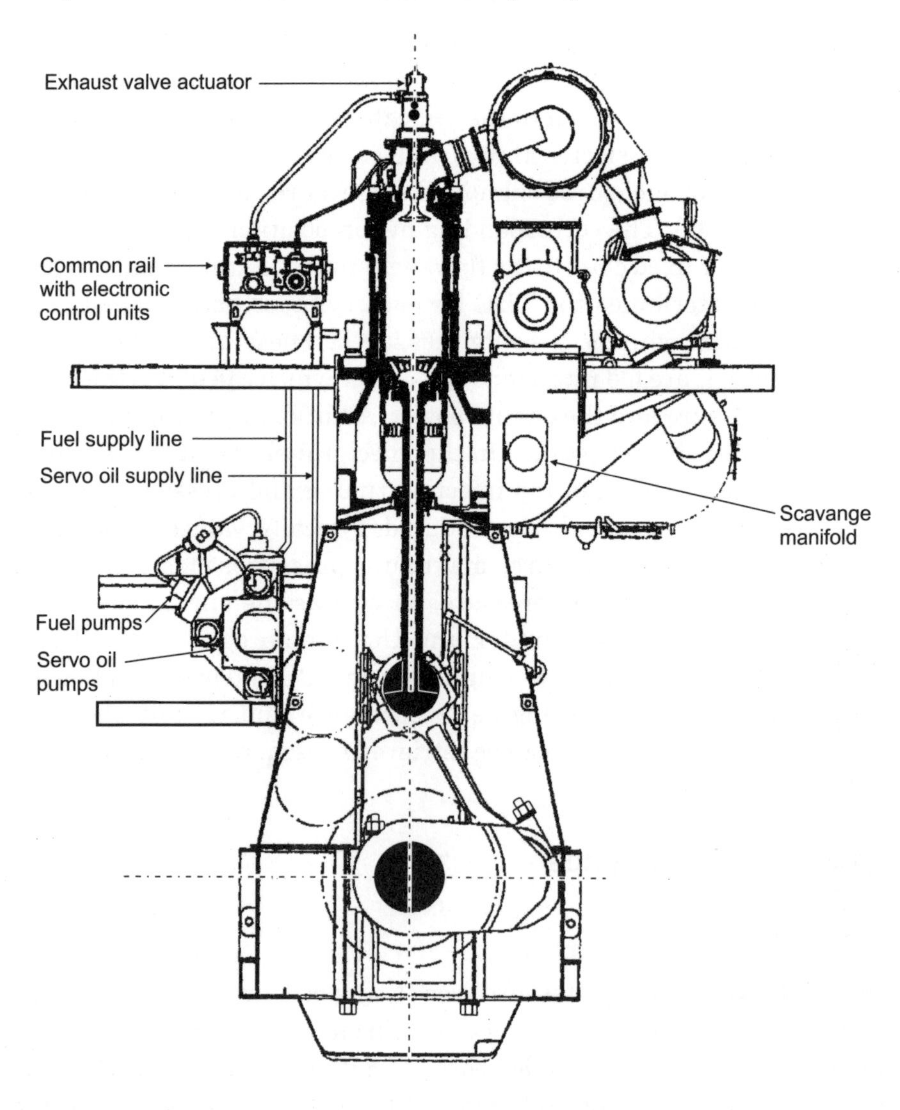

Fig. 2.9.3 Sulzer RT-flex Engine.

Sulzer RT-flex 96C electronically-controlled common-rail, turbocharged two-stroke engine is the most powerful common-rail diesel engines built so far. Developed by Wärtsilä Corporation, the Sulzer RT-flex 96C engine is currently (2009) considered the largest in the world, designed for large container ships. It is 13.5m high, 27m long and weighs over 2300 tons. This 14-cylinder engine produces more than 109,000 hp. The Electronic Engine Control Unit (ECU), which is part of a fully integrated computer controlled system, controls the following functions:

- Starting air valves
- Start and reversing
- Auxiliary scavenge air blowers
- Electronically profiled injection control
- Exhaust valve actuation
- Governor function
- Cylinder lubrication
- Engine protection

Each of the above functions of the RT-Flex engine (Fig. 2.9.3) is explained in detail under corresponding sections in chapter 4. A brief description is given here.

While the earlier jerk-type fuel injection system combined pressure generation, timing and metering in the injection pump with only limited control over the variables, the common-rail system on the other hand separates the various functions of the fuel injection system offering more flexibility in the control of the parameters, like precise volumetric fuel injection control, variable rate of injection and complete control of the injection pressure. The system employs high-efficiency fuel pumps that run on multi-lobe cams to deliver adequate quantities of fuel to the common rail at a high pressure ready for injection, the common rail being a manifold running the length of the engine near the cylinder heads. It not only provides a storage volume for the fuel oil but also dampens dynamic pressure waves. Fuel is delivered from the common rail through injection control units (ICU) to fuel injectors. The individual control unit for each cylinder regulates injection timing, controls the volume of fuel injected and also determines the spray pattern. There are generally three fuel injectors on each cylinder head, arranged around the exhaust valve, and are independently controlled so that each injector can be programmed to operate separately or together depending on running conditions. The system is designed specifically to work on heavy fuel oil.

RT-flex engines also employ hydro-electronic exhaust valve actuation and starting air control. Exhaust valves are actuated in almost the same way as in Sulzer RTA engines with a hydraulic pushrod, with the difference that the actuating energy comes from a servo oil rail maintained at 200 bar pressure. The same gear-drive that drives the fuel pump also drives the high-pressure hydraulic pump that pressurises the servo oil. The electronically-controlled actuating unit for each cylinder gives full control over exhaust valve opening and closing patterns. As part of the Electronic Engine Control Unit (ECU), each cylinder unit has its own Cylinder Control Unit (CCU) that controls cylinder lubrication through a pulse lubrication system. Similarly digital governors employing electro-hydraulic actuators ensure precise control of engine speed at varying loads.

M A N B&W MC engines generally have between six and twelve cylinders and best adapted for large ships. These crosshead-type, two-stroke engines typically have a bore of 900 mm and stroke of 2300 mm with an operating speed of 104 rpm. The slightly lower stroke translates into increased running speed. The cylinder head mounted exhaust valves are operated hydraulically under oil pressure from cam-timed actuating pistons. Constant pressure turbochargers are augmented by auxiliary blowers that kick in at low engine rpm. These are an improvement over the earlier KEF series, which were uni-flow scavenged and had a pulse exhaust system requiring grouping of cylinder exhausts into sets of three or four.

Four-stroke

Sulzer ZA series are trunk-piston engines with typical bore of 400 mm and stroke of 560 mm with a normal working speed of 500 – 510 rpm. Cylinder numbers range from six to nine for in-line and twelve to eighteen for V-engines. They employ two inlet and two exhaust valves per cylinder without valve cages. The valve seats are pressed directly on to the cylinder head. A unique feature of these engines is their rotating pistons, actuated by the tip of a hinged tongue that engages the notches of a ratchet ring inside the piston. A steel sphere at the top of the connecting rod works inside a bronze bearing cage inside the piston. The rotation ensures even temperature of the piston crown allowing reduced wear and improved piston ring and liner lubrication.

Wartsila Vasa engines are medium speed four-stroke trunk piston engines that operate efficiently on low grade fuel oil, facilitated by the use of twin fuel injectors. Cylinder bore of 460 mm and stroke of 580 mm are common and are designed for speeds of 450, 500 and 514 rpm. In-line engines carry between four and nine cylinders while V-engines have between twelve and sixteen cylinders. The twin injector system employs a pilot injector placed in a pocket at the front of the cylinder head angled down at 45° and the main injector positioned inside a central pocket. The cam-driven fuel pump has helix control and supplies the pilot charge at lower pressure and earlier than the subsequent main charge. This facilitates easy starting and efficient combustion of the low grade fuel.

M A N – B&W L series have typical cylinder bore of 580 mm and stroke of 640 mm operating at a speed of 400 rpm, making them one of the largest in-line medium-speed four-stroke engines. Cylinder numbers range from six to nine. Cylinder heads carry two inlet and two exhaust valves housed in cages. Constant pressure turbo charger supplies adequate charge air over the full range of engine load. Cam-driven helix-controlled fuel pump supplies the fuel injector, mounted centrally on the cylinder head. They operate well on heavy fuel.

❑

Chapter **3**

Engine Components

Bedplate, Frame and Tie-bolts

The bedplate provides a rigid support for the main bearings and crankshaft. It forms the foundation on which the engine is mounted and is subjected to heavy fluctuating loads from the pistons and the rotating crankshaft. It also transmits the propeller thrust to the hull structure. The bedplate therefore ensures the alignment of not just the cylinders and pistons but the propeller shaft as well.

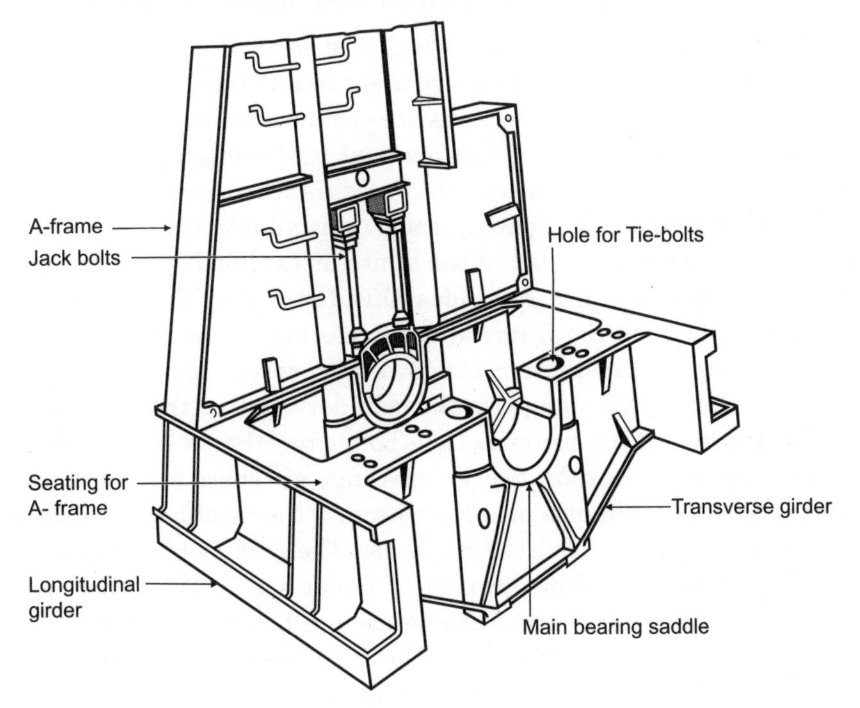

Fig. 3.1 Bedplate and frame.

The bedplate is fabricated from steel plates and comprises of two box-type longitudinal girders joined together by number of transverse pieces, one for each main bearing and one each at the two ends. It may also be made of cast-iron sections bolted together. Modern large two-stroke engines use a combination of cast steel and fabricated steel. Welds are thoroughly inspected for cracks and sub-surface flaws using the dye-penetrant method or ultrasonic devices. Next, the bedplate is stress relieved by heating in a furnace and slow cooling. While stiffeners and webs add rigidity, access holes help reduce the overall weight. Transverse girders support a steel saddle at their mid sections to carry the main bearings. The transverse saddle supports are of solid steel forgings or castings and are double butt-welded (after stress relieving by annealing) to the longitudinal girders. They are strong enough to withstand the bending moments set up by the tension in the tie-rods and the downward firing load. Holes either side of this casting allow tie-bolts to pass through. To prevent excessive bending moments in the transverse girders the tie-rod centres are closely pitched. Since this does not allow sufficient space, the upper halves of the main bearing shells are secured using jack bolts wedged against the engine frame. A sheet metal tray welded to the underside of the bedplate collects the draining lube oil and empties it into the drain tank below through a strainer at each crank-pit.

The frames are mounted on the bedplate and bolted together to form a strong, rigid column, which in turn is secured to the bedplate by fitted bolts. The individual frame sections are symmetrical in shape and carry the crosshead guides, which are made of cast-iron and bolted to the frame. Shims placed in between allow clearance adjustment (max 0.7 mm between guide and slippers). Lube oil is admitted through drilled holes and channels provided in the white-metal lined crosshead guide slippers that bear against the guides. On either side of the engine, large rectangular doors in the frame allow access to the crankcase while on the fuel pump side inspection windows with covers are provided. The column or the A-frame supports the cylinder block or the entablature, formed of individual cast iron sections bolted together, which houses the cylinders. It also incorporates the scavenge air space and the cooling water spaces. The diaphragm at the bottom of the entablature separates the scavenge air spaces from the crankcase. The cylinder block is provided with long studs for holding down the cylinder heads. The under side of the block and the top of the A-frame are machined, then aligned and bolted together using fitted bolts. The function of these bolts is only alignment and location of the three components: the bedplate, A-frame and the entablature. It is the job of the vertical tie-bolts passing through these components to hold them all together and transmit the combustion loads (cyclic stress) from the cylinder heads to the bedplate. Main bearing jack bolts are slackened off before tightening the tie-rods, which is done hydraulically to the correct pre-tension so that the engine structure is always under compression. After the tie-rods are stressed the jack bolts are tightened. Pinch bolts prevent vibration of the long tie-bolts by restricting lateral movement.

Correct alignment of the engine mating surfaces and correct tightening of tie-bolts is crucial as any error in this area could result in fatigue fracture and breakage of tie-rods. In addition, loose tie-bolts could cause fitted bolts that hold the bedplate, frame and cylinder jackets in alignment also to stretch and brake. Further, the transverse girders holding the main bearing could bend and result in cracks and main bearing misalignment. Conversely, an over tightened tie-rod could lead to its breakage in service.

Bedplate Seating

It is crucial that the bedplate be rigidly secured to the engine structure so that the alignment of the crankshaft is maintained. The engine is mounted on a foundation on the top surface of a tank. The foundation is carefully designed to be stiff and rigid. Both the bedplate and the tank top surface that receives the holding down bolts are accurately machined to be perfectly parallel. Direct drive engines are aligned accurately with the aid of laser beams to bring the main bearing centreline in line with the propeller shaft, jacking up the bedplate as required. Bolt holes are then drilled on the tank top and tapped. Next, cast iron or steel chokes are inserted between the jacked up engine bedplate and the tank top. The chokes are accurately machined to obtain a perfect fit in the gap and are individually fitted. A bolt passes through each choke, but additional chokes without bolts may be provided to carry the weight of the engine. After ensuring that all chokes are tight under the bedplate the bolts are hardened down. O-rings or other seals are used with the bolts to ensure water tightness of the tank.

It is to be noted that fitted bolts through cast-iron chokes do not serve the purpose of locating the bedplate in place. Chokes placed at the sides and ends of the bedplate and provided with through-bolts fulfil this function, preventing lateral or end movement of the engine. Improperly fitted chokes could lead to eventual fretting of choke surfaces and broken holding-down bolts, followed later by fretting of tank top surfaces.

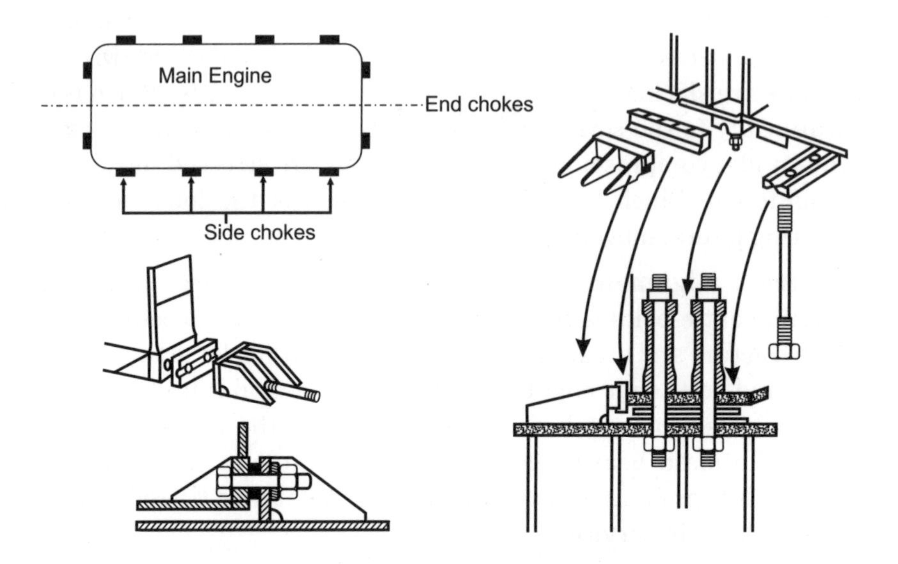

Fig. 3.2 (a) Bedplate secured with metallic chokes.

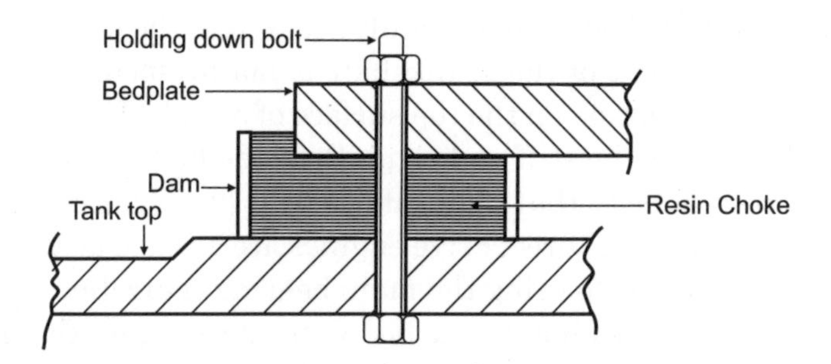

Fig. 3.2 (b) Bed plate secured with epoxy resin choke.

In an alternate method, epoxy resin, an adhesive, is used as the choking material. This offers saving in the time required to prepare the foundation for mounting the engine, since no machining of the mating surfaces is required. Once the engine is correctly aligned with the propeller shaft, the bedplate and tank top seating surfaces are thoroughly cleaned with solvents to remove traces of oil, paint and other impurities. Holding down bolts are now inserted after coating them with release agents to prevent the epoxy from adhering. Thin sections, also coated with the release agent, are now secured to the outer side of the bedplate fashioning a dam around the holding down bolts to facilitate pouring the resin and also allowing a small head above the bedplate. After a final inspection of the engine alignment, the resin and activator are mixed in a special equipment and poured into the dammed sections around the bolts. Curing takes place in about 20 to 48 hours depending on the ambient temperature. Because of the very close fit between the resin chokes and the bed plate, chances of loosing of the holding down bolts is greatly reduced, as also overall stresses acting on the bolts. Other advantages are, it is comparatively cheaper and requires less installation skills, ensures complete contact even if surfaces are uneven and unlike metal, resin is non-corrosive. However, care must be taken not to over tighten the bolts because the resin can shatter if over stressed.

Maintenance involves periodic inspection of the bedplate for evidence of cracking. Cracks could occur at the welds joining the transverse members with the longitudinal girders and below the bearing saddles. Cracks could be either radial or follow the line of the bearing pocket and could result due to number of reasons: engine overloading, loose tie-bolts or loose holding down bolts. Metal chokes must be hammer tested at periodic intervals, especially after heavy weather or damage.

Cylinder Head

The cylinder head sits over the cylinder liner flange and seals against the combustion chamber. A copper joint ring is placed in a groove in the liner flange and the head is held down against it by nuts engaging long studs screwed into the cylinder block. This transfers the gas loads acting on the head to the cylinder block from which the tie-bolts transfer it to the bedplate and then to the ship structure. The head carries all the valves required to operate the engine: fuel injector, air-starting valve, relief valve, indicator cock and the exhaust valve. The last indicated would be a centrally located hydraulically operated valve held in a cage, in the case of large two-stroke engines with uni-flo scavenging and a pair of cam-operated, push-rod actuated poppet valves for a medium speed four-stroke engine.

The cylinder head is generally of an intricate design and of considerable strength to withstand the heavy gas load and the consequent bending stresses. It has to resist the heavily fluctuating loads and transfer the heat of combustion readily. It should also be symmetrical in shape and have the same coefficient of thermal expansion as the liner.

Large two-stroke engines earlier had cylinder heads in two parts, both fresh water cooled. The lower part was of cast steel for strength and held against the liner flange by long studs. The central upper part was a spheroidal graphite cast iron insert held down on to the lower part against a soft iron joint ring by studs. It carried the fuel injector, starting air valve, the relief valve and the indicator cock. The insert was an exceptionally stiff casting while the lower part was thin-walled to ensure effective heat transfer with a minimum of thermal stresses. Cooling water was fed tangentially (to avoid impinging on the opposite surface and avoid erosion) into the lower part from where it passed into the central insert. Later engines however, had single-piece forged steel cylinder heads. These employed 'bore cooling' wherein series of tangential holes were bored through the head; all linked at intersections, and carried the water very close to the combustion chamber ensuring more efficient cooling. It is usual to have the cooling water for the head in series with the jacket. Coolant exiting from the cylinder jacket top is directed into the base of the cylinder head, the increased water temperature compatible with the higher temperature of the head, thus reducing thermal stressing.

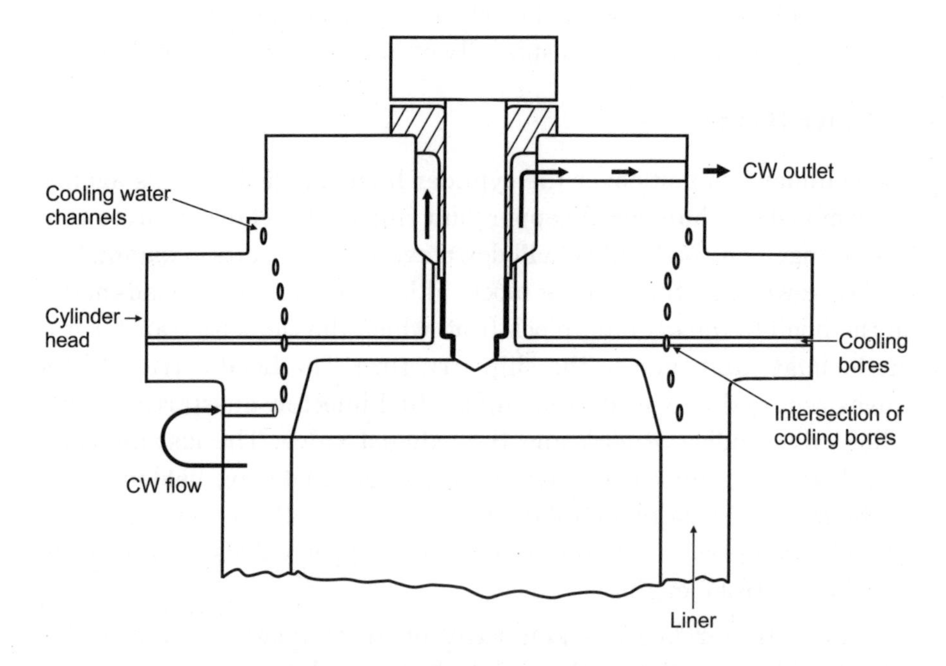

Fig. 3.3 Single-piece cylinder head.

Two-stroke Engine Hydraulically Operated Exhaust Valve

In modern uni-flow engines, the hydraulically operated exhaust valve, centrally located in the cylinder head, sits in a valve cage that is bore cooled, ensuring effective cooling of the cage and valve seat. The flat piston crown used in these engines necessitates a shaped head in order to form the combustion chamber, making the design complicated.

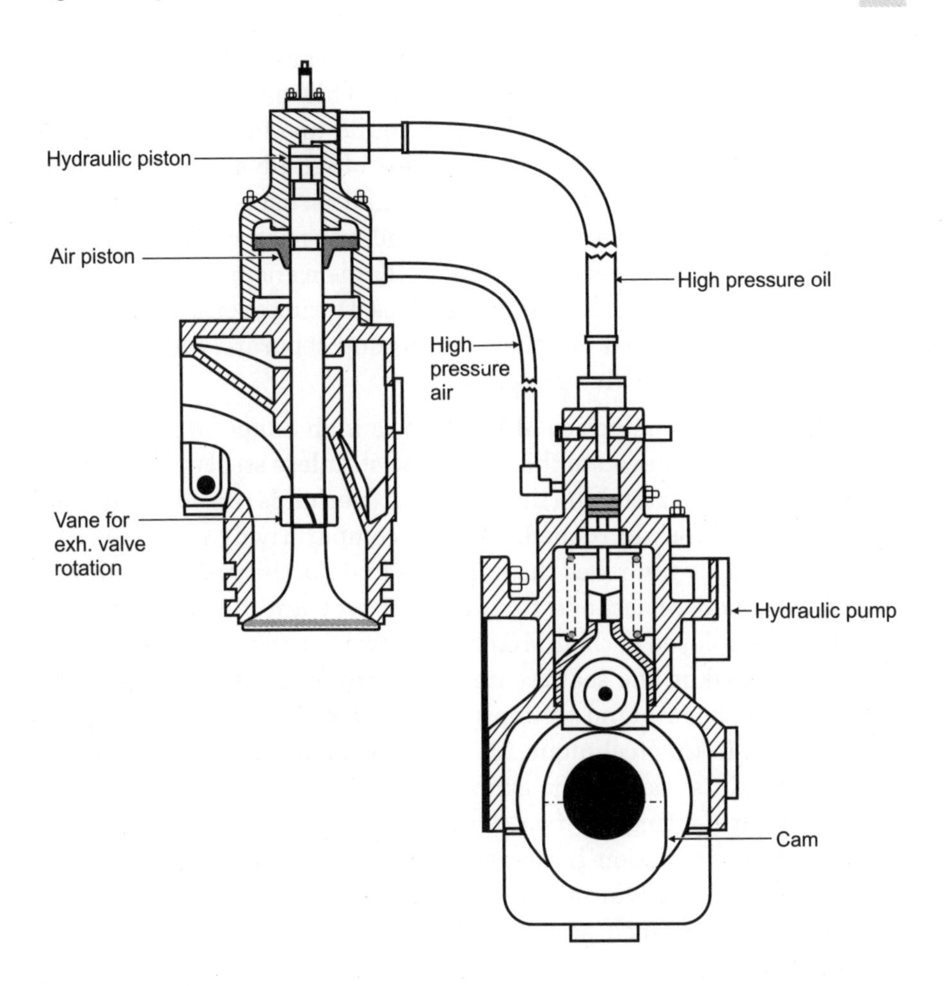

Fig. 3.4 Hydraulically operated exhaust valve for large two-stroke engine.

Fig. 3.4 shows the working of such a valve for a M A N B&W two-stroke slow-speed engine. The cam operates a hydraulic pump (instead of a push rod) and the delivered oil (taken from the engine lube oil system) acts on a piston in the exhaust valve that pushes the valve down to open it. In place of helical springs that close a mechanical valve, here an 'air spring' is used. High-pressure air is led via a non-return valve to the underside of a piston connected to the valve spindle. As the piston moves downward during opening, it compresses the air further and when the hydraulic pressure on the valve is relieved the compressed air pushes up the valve to close it. To prevent air lock an air vent is provided at the top of the pressure oil chamber. Oil leaking from this vent is made up through an oil make up line connected to the oil pump cylinder. A dampening arrangement atop the valve piston prevents hammering of the valve on the seating. The valve spindle is provided with a winged valve rotator. Escaping exhaust gases rotate the valve a small amount as they pass over the rotator. This maintains the valve at an even temperature and also prevents build up of carbon deposits on the valve seat. The hydraulic system offers other advantages too: in the absence of transverse thrust, the purely axial thrust acting on the valve reduces guide wear, controlled valve landing speed reduces stress on the valve seat and the valve rotation mechanism ensures uniform valve seating and more balanced thermal stresses. During overhauls, the valve and seat are not lapped together, but ground separately to the correct angles using special valve grinders.

In a new development by MAN B&W, a high temperature-resistant Ni-Cr alloy is welding on to the seat of the stainless steel spindle, which has been found to extend the time between overhauls considerably when the surface is work-hardened. Though comparatively expensive, the technique greatly improves the hardness and ductility of the seat, with much improved resistance to cracks. In yet another cost-effective innovation employed in their MC engines, two narrow concentric contact areas machined on the seat create a new seat geometry, called the W-seat that has been found to improve the overall condition of the seat contact area. In this technique, all the carbon particles caught between the valve and its seat on the cylinder head get crushed and squeezed out – without denting the two contact areas. Made of solid, low-alloy steel, with an induction hardened seating surface, the W-seat offers considerable cost advantage as well as significantly increasing the working life – as much as 40,000 running hours without needing to grind the seat.

Four-stroke engine cylinder heads have more number of opening to accommodate a pair each of inlet and exhaust valves in addition to the central fuel injector, starting air valve and the relief valve. Inlet air and exhaust gases are led through internal ducts integral to the head. Spheroidal graphite cast iron or in some cases cast steel is used for the box-like main head casting and a steel flame plate is attached to the bottom that also carries the valve seats. The seats are shrunk-in replaceable inserts made of wear-resisting molybdenum cast-iron or a cobalt-chromium alloy.

Cooling passages inside the head reach close to the seats to ensure effective cooling (Fig. 3.5). Exhaust valve and spindle is a forging of high-tensile heat-resistant steel. Closing of the valves is ensured by a set of helical springs placed around the spindle between two spring plates. The head is secured to the liner by four high-tensile steel studs screwed into the engine frame; and gas sealing is achieved by a copper packing ring inserted between the head and the liner. Rocker arms mounted on the head and actuated by push rods operate the inlet and exhaust valves. Each rocker arm, oscillating on a fixed hardened steel shaft, actuates a cross head that rests across a pair of inlet/exhaust valve stems. A clearance is maintained between the valve stem and the rocker arm end when the engine is cold, which is taken up by expansion when the engine attains normal running temperature.

During maintenance these clearances are adjusted carefully since too little clearance will result in incomplete closing of the valve while too great a clearance can result in reduced valve lift.

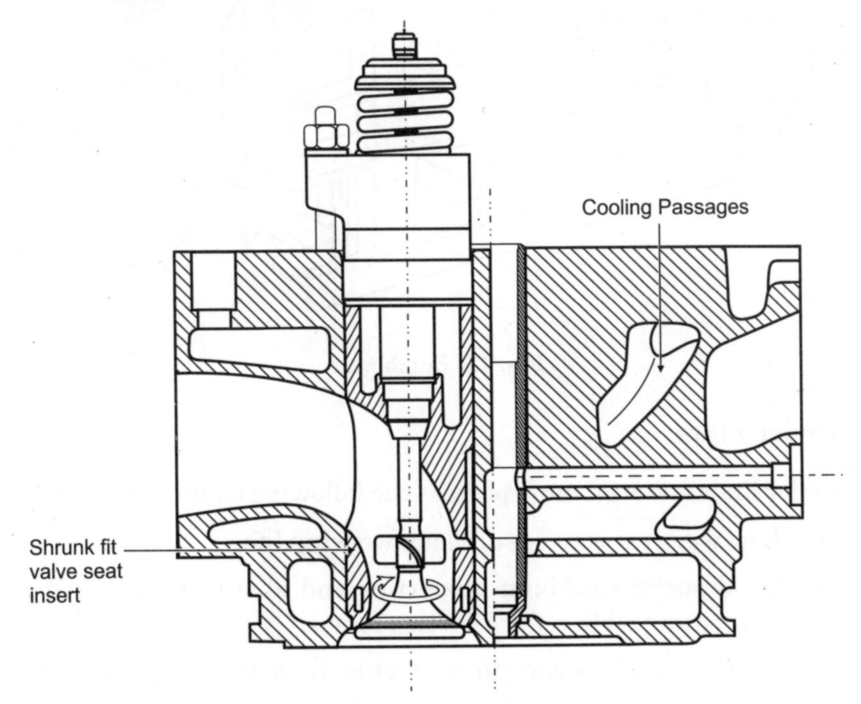

Fig. 3.5 Four-stroke engine cylinder head – M A N B&W.

Entablature

This is a structure formed of individual castings bolted together to form the cylinder block which incorporates the scavenge air space and the cooling water spaces as shown in Fig. 3.6. It forms the housing to take the cylinder liner and is made of cast iron. It may also be cast in multi-cylinder units, which are then bolted together to form a beam. The underside of this beam is machined and aligned on the A-frames and fastened in position using fitted bolts. Long bolts threaded on to its top side are used to hold down the cylinder heads. In large slow speed two stroke engines, the entablature may hold only the scavenge spaces and the stuffing boxes, while a separate cylinder block made of individual castings bolted together is mounted on top of the entablature.

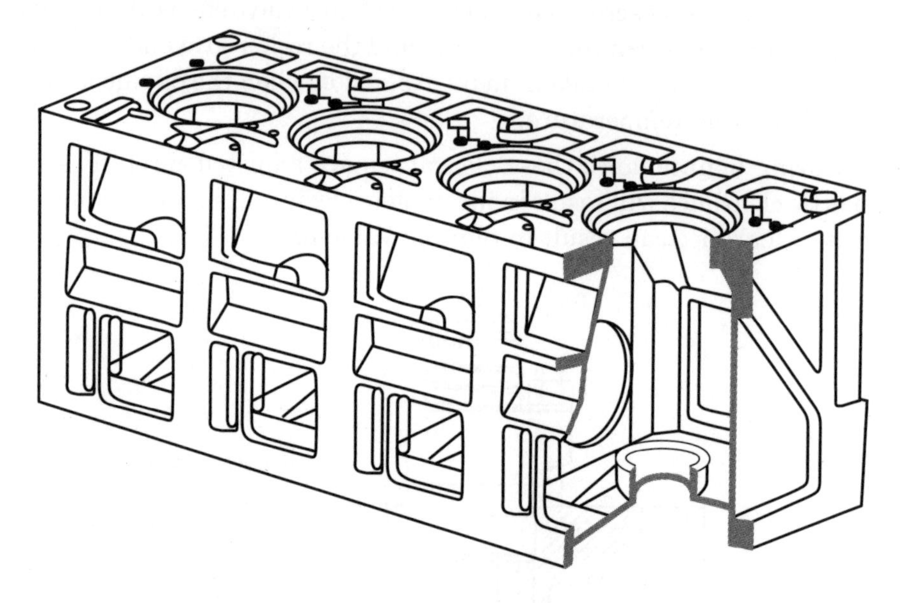

Fig. 3.6 Entablature.

Cylinder Liner

Cylinder liner material must possess the following characteristics:

- Low wear rates to ensure long working life.
- Good surface rubbing properties and able to retain a film of lubricating oil to reduce friction losses.
- Possess high strength and able to withstand fluctuating pressures, temperatures and stresses in order to ensure good fatigue life.
- Good conductor of heat to readily transfer heat of combustion to the circulating cooling water.
- Resistant to abrasion and corrosion.
- Have a rate of thermal expansion compatible with mating components.

Cylinder liners are commonly made from pearlitic grey cast iron because of the relative ease of casting, good wearing characteristics and self-lubricating properties due to the graphite content. Mechanical strength is enhanced by creating special alloys with vanadium and titanium. A porous deposit of chrome is plated on to the inner surface in some engines to reduce wear and withstand corrosion better, thus enhancing life. The porous structure helps in retaining lubrication. However, in this case the piston rings must not be chrome plated, for that will wear out the deposit fast. Fig. 3.7 (a) and (b) show liners of two-stroke and four-stroke engines respectively.

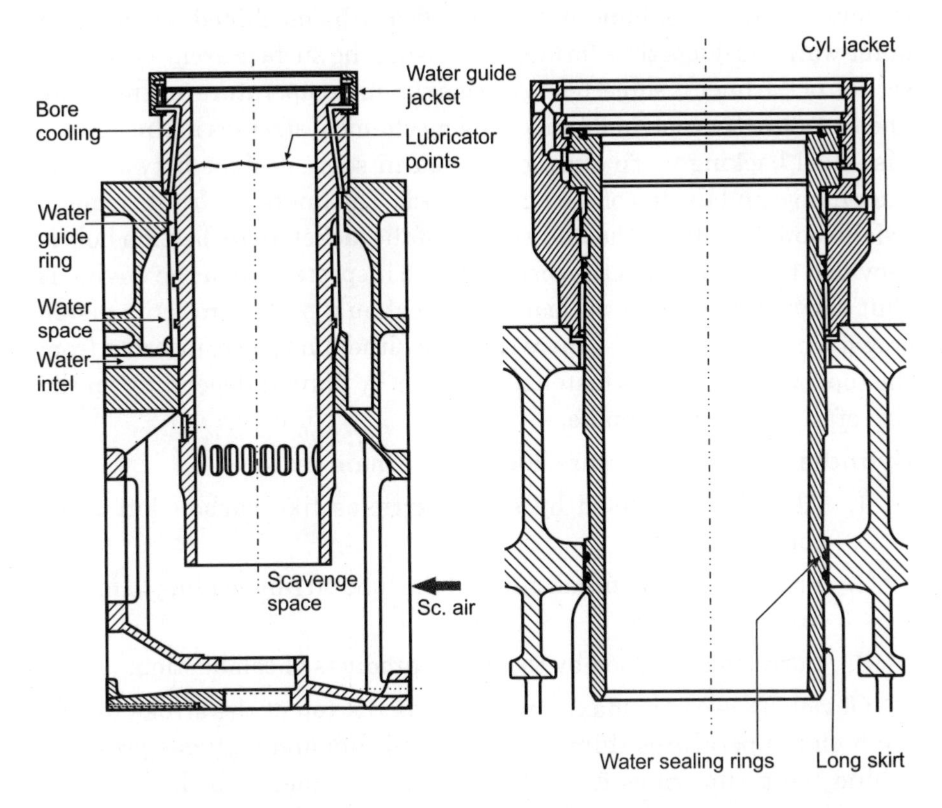

Fig. 3.7 (a) Cylinder liner – 2-stroke engine. (b) Cylinder liner – 4-stroke engine.

Liners with scavenge and exhaust ports machined in are fitted inside jackets formed by bolting together number of similar iron castings that constitute the rigid engine block. The block rests on the engine frame and held in compression by tie-bolts. Port edges are shaped to direct the flow of air and gases to obtain maximum scavenging efficiency. The flange at the top end of the liner rests on top of the block. Cooling water circulates through the annular space formed between the liner and the block. The spaces above and below the ports and at the top of the liner are sealed against cooling water using high-temperature resistant neoprene rubber O-rings fitted into grooves in the liner. An empty inspection groove is machined in between the set of twin O-rings and connected to the outside of the engine through a hole in the liner. This allows easy detection of water or gas leakage. The thin-walled but strong-backed design of the liner ensures efficient heat transfer to the circulating cooling water, at the same time imparting the required strength to withstand the high temperature and pressure conditions, without causing undue thermal stresses and possibility of cracking. Cooling is further enhanced by the helical ribs machined on the liner outer wall; the ducts thus formed increasing the surface area. The upper section of the liner is subjected to the maximum temperature and pressure and to reduce thermal and tangential mechanical stresses in this area, a cast-steel backing or **fire ring** is fitted in some designs between the liner flange and the top of the engine block. The liners are free to expand axially downwards. In the mid-section of the jacket an inspection hole is provided through which the piston can be inspected when the engine is shut down and also allows cleaning the exhaust ports across the piston crown when it is at BDC. Liners could be either one piece, inserted from the top, or in two pieces fastened together by flanges depending on the size of the engine and make.

Cylinder liner wear occurs due to three main reasons:

1. Abrasion – caused by solid particles like carbon left after combustion.
2. Friction – due to break down of the lube oil film leading to metal-to-metal contact.
3. Corrosion – caused by the acidic products of combustion.

Obviously, wear is maximum towards the top of the stroke due to the high temperatures thinning out the oil film and high gas pressure behind the piston rings forcing them into the liner wall. In addition, since piston speed is lowest as it nears the end of its stroke a good oil wedge cannot be formed. Wear rates reduce lower down the stroke because pressure and temperature conditions are less severe here. At the bottom end of the stroke wear rate increases again due to reduced piston speed and the scouring effect from the incoming scavenge air. The reduced temperature also increases the viscosity of the oil weakening its ability to spread evenly.

Another cause of liner wear is the accumulation of carbon on the piston crown. Build up of carbon deposits on the edges of the crown can eventually begin to rub against the liner producing a hardened glassy surface on the rings and liner. This **glazing** renders the surfaces incapable of holding an oil film. To prevent this, an **anti-polishing ring** is fitted in some designs – a clearance fit inside a step at the top of the liner. The ring has a slightly lower inner diameter than the liner and only marginally larger than the piston crown. As the piston reciprocates, the ring scrapes off carbon deposits from the edge of the crown.

Finally, acids formed by the products of combustion cause corrosion, which is of greater consequence than frictional wear of liners. During combustion the sulphur content of the fuel gets converted to sulphuric acid which can condense on liner walls which are at relatively lower temperature. Corrosive wear can be sometimes as high as 0.4 mm per 1000 running hours. Special additives in the cylinder lubricating oil maintain an alkaline ph to neutralise the acids. It is clear therefore that lubricating oil quantity is governed also by the engine bhp or in other words on the quantity of fuel burned, and its sulphur content.

Cylinder Lubrication

The cylinder liner is lubricated by the Pulse Jet Lubricating System in which multiple quills positioned around the liner inject oil, supplied by a lubricator pump, into grooves in the liner wall that spread and retain the oil film, Fig. 3.8.

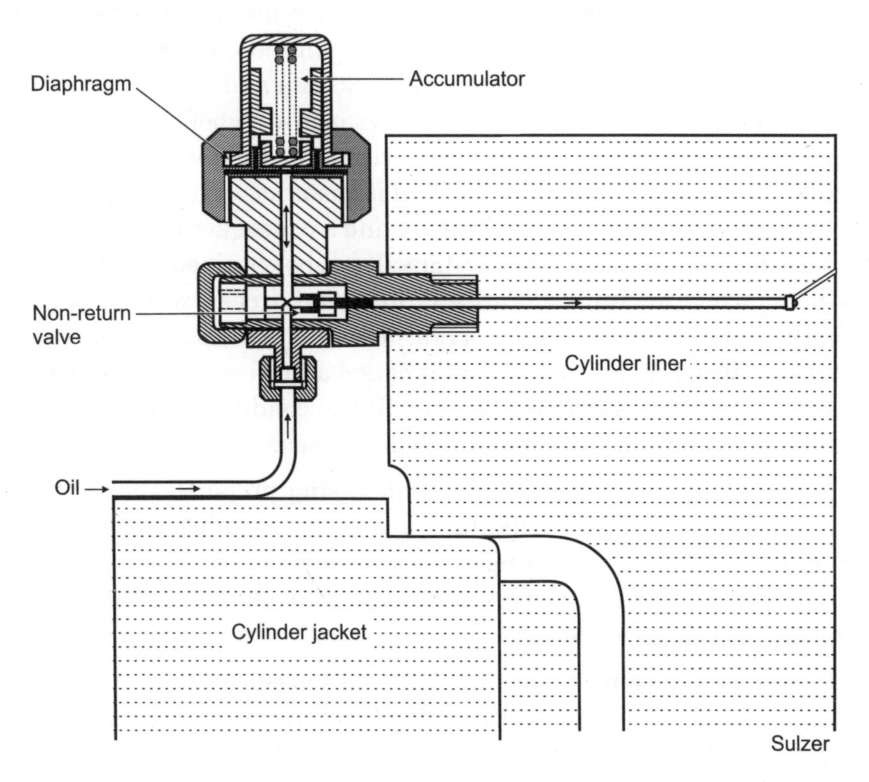

Fig. 3.8 Lube oil quill fitted to a liner.

The groups of pumps, made up of blocks of individual lubricators, are actuated by an oscillating shaft which is connected to a crank at the aft end of the camshaft. The oscillating shaft is fitted with drive levers, one for each cylinder that allows adjustment to be made to the feed rate of individual cylinder that can be checked by the position of balls through a glass tube. The feed rate is based on various factors like the type of cylinder oil used, type of fuel, engine load and condition of piston and liner, and varies from 0.3 to 0.6 gm/bhp/hr. The quantity of oil delivered to the liner cannot be proportional to the speed of the engine alone, since at partial or low loads, the quantity of fuel burned is reduced, hence to prevent excessive lubrication cylinder oil delivered must also be reduced. This is achieved by variable adjustment of the lubricators obtained by adjusting the angular movement of the lubricator crank.

Injection timing is critical and if timed precisely to inject oil between the piston rings as they pass, the oil is spread upwards evenly by the piston rings. And since it is when the piston is leaving the BDC that it is at its slowest and also subjected to the least pressure, this is the ideal instant when lubricating oil is injected from the feeding points. A non-return valve is invariably provided in the oil delivery line before the quills to prevent blowback. The number of quills is governed by the diameter of the liner, and M A N B&W engines are provided with 8 feeding points for liner diameter up to 780 mm and 10 or 12 points for higher diameters. Lubricators are always operated before starting the engine to prime the lines and to pre-lubricate the liner walls for the first engine movement. If the quantity of oil is insufficient (or has a reduced viscosity due to high liner temperature and a consequent thinning of oil film) a complete and effective oil film will not be obtained between liner and piston rings leading to localised **micro-seizures** – when asperities on both surfaces begin to have metal-to-metal contact. The process is called **scuffing** and produces what is called adhesive wear, resulting in a rough, scorched surface on the liner. If the conditions are severe the rate of wear can accelerate leading to piston seizure.

If the oil has inadequate acid neutralising properties for the fuel being burnt (due to depleted alkalinity) corrosive wear in a pattern called **cloverleafing** result, leading to gas blow-by past the rings. Over lubrication can lead to sticking rings and choked ports due to accumulation of un-burnt oil.

In **modern engines**, like the Wärtsilä low-speed engine, the feed rate and timing are electronically controlled through a solenoid valve at the lubricator pump offering flexibility in setting the timing. The volumetric metering ensures constant injection patterns across the engine's load range and maintains a feed rate of 0.6 gm/kwh (0.44 gm/bhph). The system ensures low wear rates and improved liner and piston ring conditions.

The Electronically-controlled Wärtsilä Pulse Lubricating System

High price of cylinder lubricating oil and its contribution to air-polluting emissions have made it necessary to lower cylinder oil feed rates. The electronically-controlled Wärtsilä Pulse Lubricating System which is standard in large-bore Wärtsilä low-speed marine engines enables very low lubricating oil feed rates to be used without compromising lubrication expediency. A flexible and precise timing of oil delivery, accurately metered and load-dependent quantities of oil and improving the distribution of the oil on the cylinder liner surface are the other advantages. This has resulted in considerable cost savings, as much as 35 %, and improved liner life. Electronic control ensures not only accurate dosage and timing but also complete flexibility in settings. This System is incorporated in the RT-flex 96C and RTA 96C engine types, with other engine types being added since 2007.

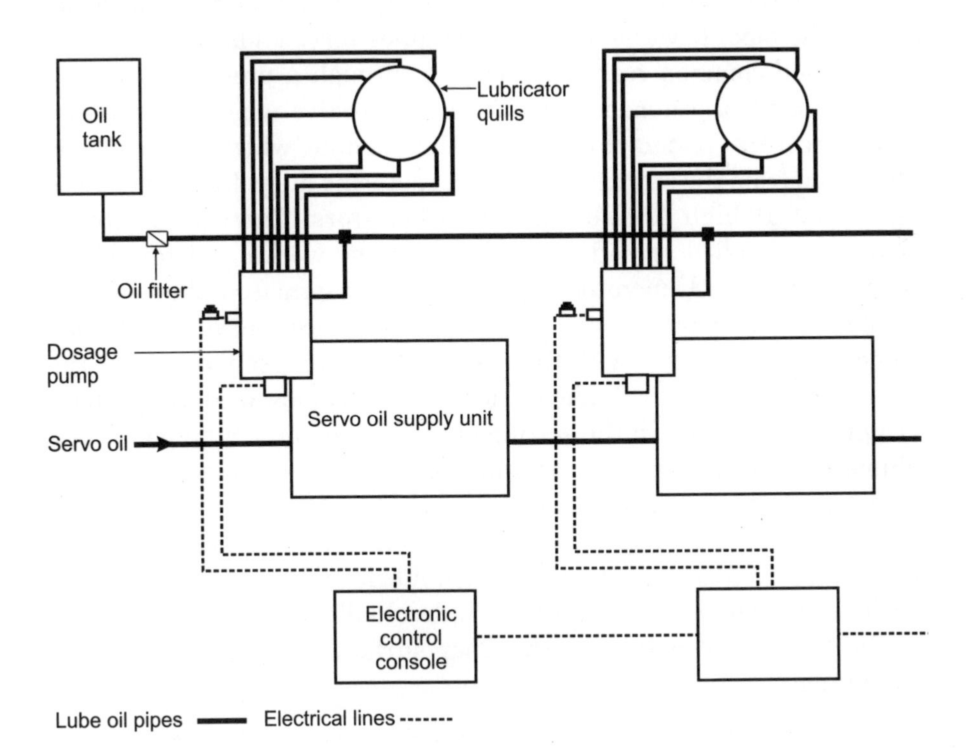

Fig. 3.9 Schematic diagram of the Pulse Lubricating System.

As shown in Fig. 3.9, the main components of the system are:

- A pulse lubricating module consisting of a dosage pump with electronically-controlled timing.
- Lubricators, arranged in a single row around the cylinder liner
- Filtering and measuring system
- Servo-oil supply unit (on RTA engines) or pressure reducing unit (on RT-flex engines)
- Electronic control system
- Crank angle sensors.

Each cylinder has its own separate lubricating module thereby allowing short delivery pipes to the lubricators. Each module consists of a dosage pump, a solenoid valve, electronic controls, a pressure sensor and a diaphragm accumulator. The dosage pump, powered by the pressurized servo-oil, delivers a metered quantity of cylinder oil at high speed to the lubricators at the precise timing ascertained by the control system. Cylinder lubricating oil and servo oil are supplied separately to each module. On RTA engines a separate servo oil supply unit is employed to drive the module while on RT-flex engines servo oil is supplied from the servo-oil common rail of the engine through a pressure reducing valve. The control system actuates the solenoid valve, which directs servo oil to the central piston of the lubricating module, which in turn feeds cylinder oil from the storage tank to the metering ducts from where it is discharged at high pressure to the lubricators. The cylinder oil is accurately supplied at defined positions of the working piston whose position is constantly monitored by the control system from the reference signal transmitted by the crank angle sensor. The electronically-controlled solenoid valve, together with the dosage pump, precisely regulates the feed rate and timing of cylinder oil delivery even at low feed rates. There is complete flexibility in the setting of the lubricator timing point, and volumetric metering ensures constant spray patterns across the engine's load range. Lubricating oil is sprayed as a pulse of multiple jets on to the liner surface from a single row of lubricators, with built-in non-return valves, arranged around the liner, each lubricator having a number of nozzle holes. The oil jet sprays are individually directed to separate, evenly distributed points on the liner surface thus ensuring better distribution.

M A N B&W ME engines employ a computer controlled cylinder lubricator with intermittent lubrication that enables a reduction in lube oil dosage of as much as 0.2 g /BHPh. Their Aplha Adaptive Cylinder

Oil Control (Alpha ACC) system employed in the MC-C engines is based on an algorithm that controls cylinder oil dosage proportional to the sulphur content in the fuel and the engine load, which in turn determines the amount of fuel entering the cylinders. Since corrosion is a major cause of cylinder liner wear, if the amount of neutralising alkaline compounds in the cylinder liner is proportional to the amount of sulphur – which generates sulphuric acid – entering the cylinders, it would afford adequate protection against corrosion. The optimal cylinder oil dosage thus determined also includes the minimum required to satisfy the other requirements of lubrication. In addition to significant savings in cylinder oil consumption, the system ensures lower particle emission levels and reduced liner wear.

Maintenance

Cylinder liners should be inspected and gauged internally at fixed intervals (6000 – 8000 running hours in older conventional engines). After the pistons are withdrawn, the liner walls and the ports should be cleaned of all carbon deposits (increasing exhaust temperatures and scavenge air pressure would indicate fouled exhaust ports). A careful surface examination will reveal any surface cracks and also if the lubrication had been adequate. A mirror like surface will indicate good running conditions. Dark, dry areas on the upper areas will indicate piston blow past while rough vertically streaked areas will indicate breakdown of lube oil film and metal seizure. The bore should be measured accurately (when the liner is cold) at number of vertical positions using a steel template in fore and aft and athwartships directions and compared with the records to determine the increase and the mean rate of wear since the last reading. In conventional engines a wear of 0.6 – 0.8 % of original bore per 1000 hours is acceptable, and 1 mm per 100 mm of diameter is the limit before liner replacement. Any ridges are smoothened and lubricators and quills tested. The cylinder jacket cooling spaces should be inspected through inspection doors provided. Scale deposits can be removed by flushing with water, but if they are too heavy, the liner may have to be withdrawn for cleaning. If the liner needs to be replaced, the rubber O-rings on its outer grooves should be ideally replaced too making sure that they do not project more than 1 mm beyond the liner wall. Jacket cooling spaces should be thoroughly cleaned. The new liner should be gauged for records and lowered into the jacket taking care to see that it is aligned circumferentially so that quill connections etc are in position. A new set of piston rings should be used with a new liner. After fitting a new liner, the engine should be 'run-in' slowly at reduced speed and temporarily increased cylinder lubrication, under preferably reduced load. This helps the asperities or high spots on the rubbing surfaces to wear off leaving a microscopic corrugated surface able to retain an oil film.

In some engines the liner can be examined without withdrawing the piston. With the piston at BDC, liner, and the exhaust ports, can be viewed through the scavenge ports, with a light introduced. A mirror will allow a clear view of the upper parts of the liner.

Piston

Though both piston and cylinder liners have to withstand high gas load, high temperatures, fluctuating thermal and mechanical stresses and stresses imposed by friction, cast iron is an unsuitable material for pistons because of its comparatively weak mechanical properties. Pistons for modern slow speed engines are manufactured from alloyed cast steel (nickel-chrome or chrome-molybdenum) or forged steel which have high strength, resist heat stresses better and withstand creep, corrosion and erosion. The improved heat conduction ensures that combustion heat (crowns reach temperatures as high as 450°C) is carried away efficiently by the cooling medium. Furthermore, due to low thermal expansion, working clearance of piston rings can be maintained. However, casting or forging processes create internal stresses and the castings are therefore **annealed** to remove or reduce these stresses as well as to refine the grain structure. Material and design of the piston depends on the engine rating, size and speed.

Medium and high-speed engines have another important consideration: the weight of the piston. This should be low in order to reduce the stresses on the rotating parts. Aluminium alloys which have high thermal conductivity and low weight make an ideal material. Cooling pipes are generally cast into the crown, which is made thicker both for strength and to aid in heat conduction. Some designs have a composite piston where the piston is made in two parts with a cast steel crown containing two grooves and an aluminium body. Thus, even though aluminium has a higher coefficient of expansion, the crown protects it from the high temperatures of combustion. M A N B&W 58/64 engines have a forged alloy steel crown of thin section bolted to a forged aluminium alloy skirt. Secondly, because aluminium has a lower coefficient of friction bushes for the gudgeon pin may be avoided, and therefore a floating gudgeon pin may be used. In four-stroke trunk-piston engines the oil scraper rings are placed away from the compression rings and lower down. Scraper rings have a sharp edge to assist in scrapping oil from the liner walls (splashed on by the rotating crankpin) back into the crankcase. The M A N engine mentioned above has three compression rings in the crown and one scraper ring in the top part of the skirt for oil control. The top compression ring is plasma coated while the lower ones are chrome plated. Plasma coating involves spraying a gas mixture on to the metal through an electric arc struck between tungsten electrodes. The extremely high temperature of the arc melts the metal surface and fuses it with gas molecules.

Worn out scraper rings increase lube oil consumption and can sometimes dictate when an engine needs overhauling.

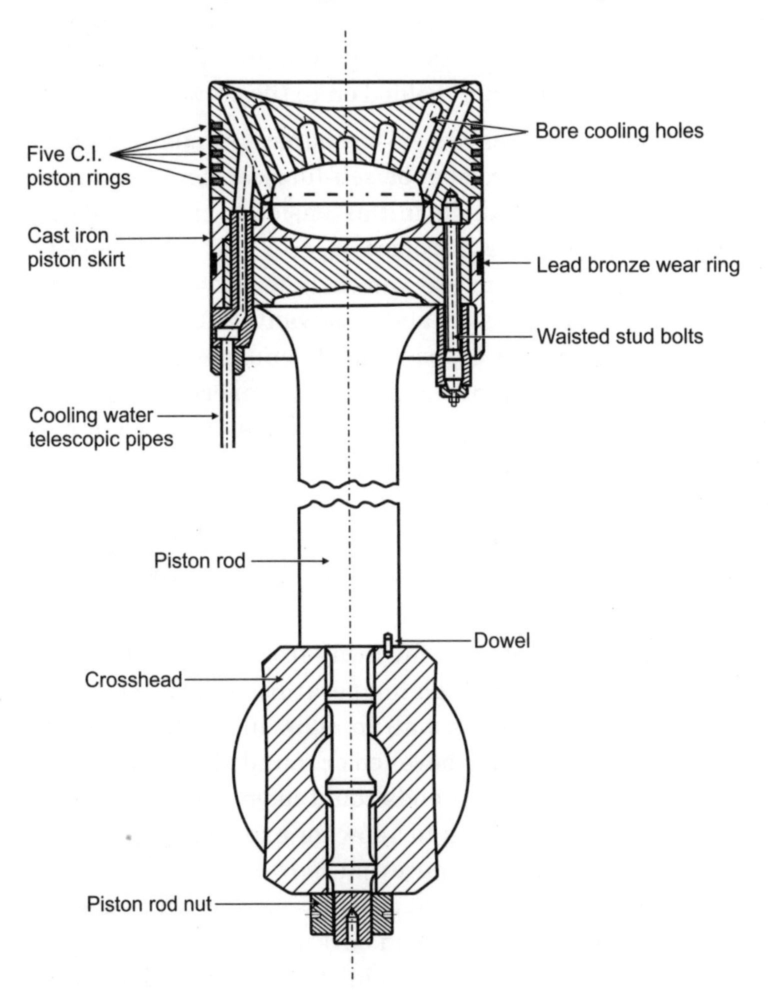

Fig. 3.9.1 Water-cooled Piston – Sulzer RTA engine.

Large two-stroke engine pistons are also made composite with a cast or forged-steel crown bolted to a skirt made from cast iron – for its better rubbing properties. Fig. 3.9.1 shows pistons of a large two-stroke engine. Even though the skirt does not have to transfer side thrust to the liner – since this is accomplished by the cross-head – it has to be long enough to cover the scavenge and exhaust ports when piston approaches TDC (in the case of cross or loop scavenging). Cast iron inserts that resist wear are fitted into the ring grooves to form a hard landing for the rings – to minimise the effects of ring hammering which can result in groove enlargement and ring breakage. The crown carries 4 or 5 grooves which may be chrome plated and carry the piston rings.

Sulzer RND engines using the loop-scavenging system have a flat crown while other makes could have either a concave shape, which offers a more efficient combustion space, or convex that has more mechanical strength. To get better protection against high temperature corrosion a protective alloy is sometimes welded on to the crown.

Piston rings should have good mechanical strength, high resistance to wear, resistance to corrosion, be self-lubricating, be compatible with the liner material and retain its initial tension to give a good gas seal. To attain a combination of these properties, grey cast iron, alloyed cast iron or spheroidal graphite iron are variously used to manufacture piston rings. Each has its advantages and disadvantages, but careful selection of alloying material from copper, molybdenum, vanadium and nickel and heat treatments like quenching and tempering help in attaining the right combination of strength, hardness and wear resistance. A protective coating of chrome plating further enhances wear properties. Tungsten carbide coating also gives better wear resistance. An outer coating of a soft metal such as copper or graphite is sometimes provided that wears quickly during running in, giving the rings a surface profile that match the liner.

The first ring is located as far away from the crown as possible to reduce ring expansion from high temperature. It is also given adequate axial clearance (twice the normal) to allow more space for expansion. The edges of all the rings are bevelled or rounded to better retain a lube oil film on the liner. The ring ends could have butt, lap or mitre joints with sufficient gap in between to allow for expansion without stressing the rings, yet small enough to prevent blow-past. The ends, which are slightly rolled in to prevent them catching the port edges, are also arranged to coincide with the broad bars at the scavenge ports. A **wear ring** made of a copper-lead-bronze alloy – having a low coefficient of friction – is generally caulked into a groove on the skirt, projecting by 1 mm, to assist in the running-in process. In conventional engines, by 5000-10000 running hours, the wear ring would be worn flush with the surface of the skirt. In medium-speed trunk-engine pistons the lead-bronze ring is proud by 0.1 mm and by the time it wears flush with the piston surface any rough running would be ironed out. The material and design of rings therefore determine whether friction and wear are the least possible.

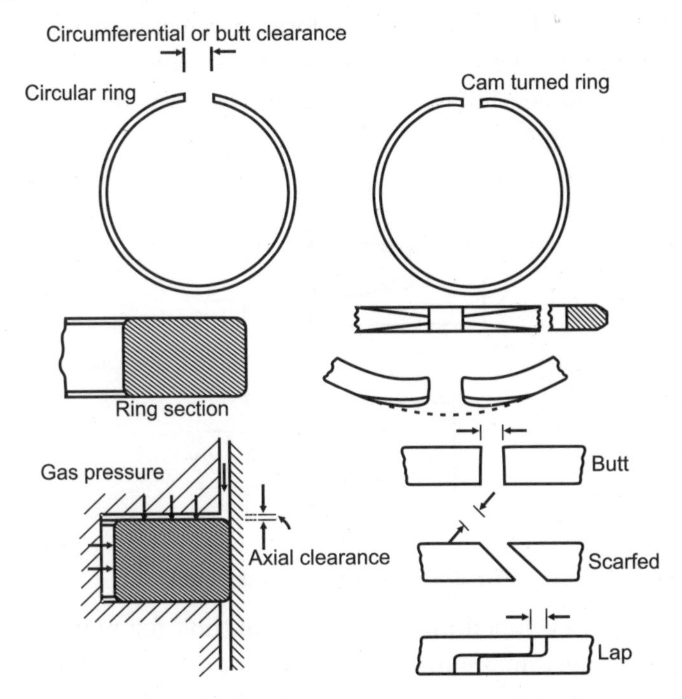

Fig. 3.9.2 Piston ring profiles. (a) Two-stroke engines.

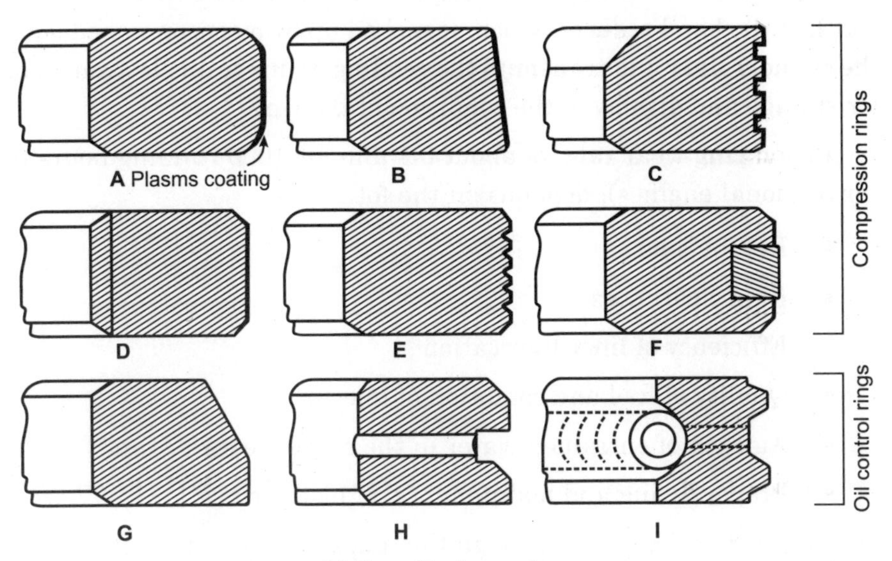

(b) Four-Stroke engines

In order to reduce running-in time M A N B&W has developed an aluminum-bronze coating for the sliding surface of piston rings in their MC engines. This reduces scuffing considerably making it possible for the normal increase in the cylinder oil feed rate after changing piston rings to be omitted. There is also considerable reduction in oil consumption during running-in when cylinder liners are changed. After about 1000 running hours the coating wears off leaving both the liner and the piston rings perfectly run-in.

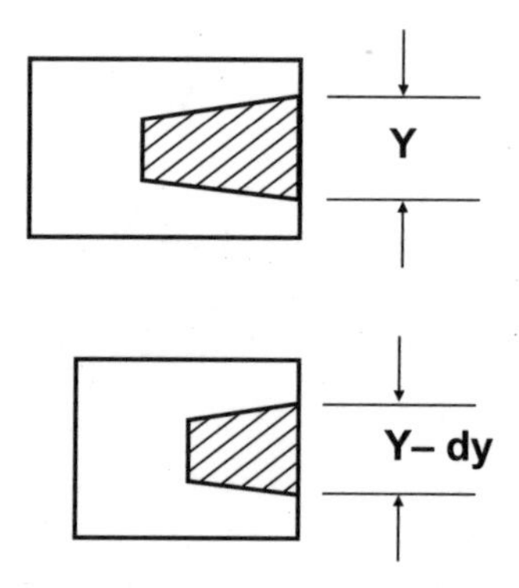

Fig. 3.9.3 Piston ring wear-band.

Sulzer has developed an innovative method for monitoring piston ring wear. The system enables a continuous online feed back of the rate of wear. The piston carries a copper wear-band embedded in its outer surface as shown in Fig. 3.9.3. As the ring wears down the exposed width of the band reduces which us picked up by a sensor embedded in the cylinder liner and transmits the reading online to an electronic unit for storage and display of the condition of the ring.

Piston ring wear rate, of about 0.1 mm for 1000 running hours (in conventional engines), depends on the following:

- Engine load
- Ring clearances
- Efficiency of liner lubrication
- Quality of fuel and condition of injectors
- Amount of entrained water in the scavenge air
- Proper fitting and maintenance of piston rings

Coolant is circulated through the inner spaces of the piston crown to carry away the heat of combustion and limit thermal stressing. Coolant could be either water or oil. Circulating water is force-fed into the crown through a telescopic pipe – fitted to the inside of the piston crown and sealed over a fixed pipe fitted through the diaphragm to a casing below it. As the piston reciprocates, water inside the crown along with the air present gives rise to a shaker effect, further enhancing the cooling effect. In the case of **oil-cooled engines,** cooling oil forms part of the engine lube oil system and is first supplied to the cross head through either telescopic pipes or articulated pipes that swing as the crosshead reciprocates. From here, oil passes through a central passage bored inside the piston rod to the cooling spaces below the piston crown. As shown in Fig. 3.9.4 oil returns through an outer passage concentric with the inner bore down the piston rod and collects in a receiver before draining out into the crankcase for recirculation.

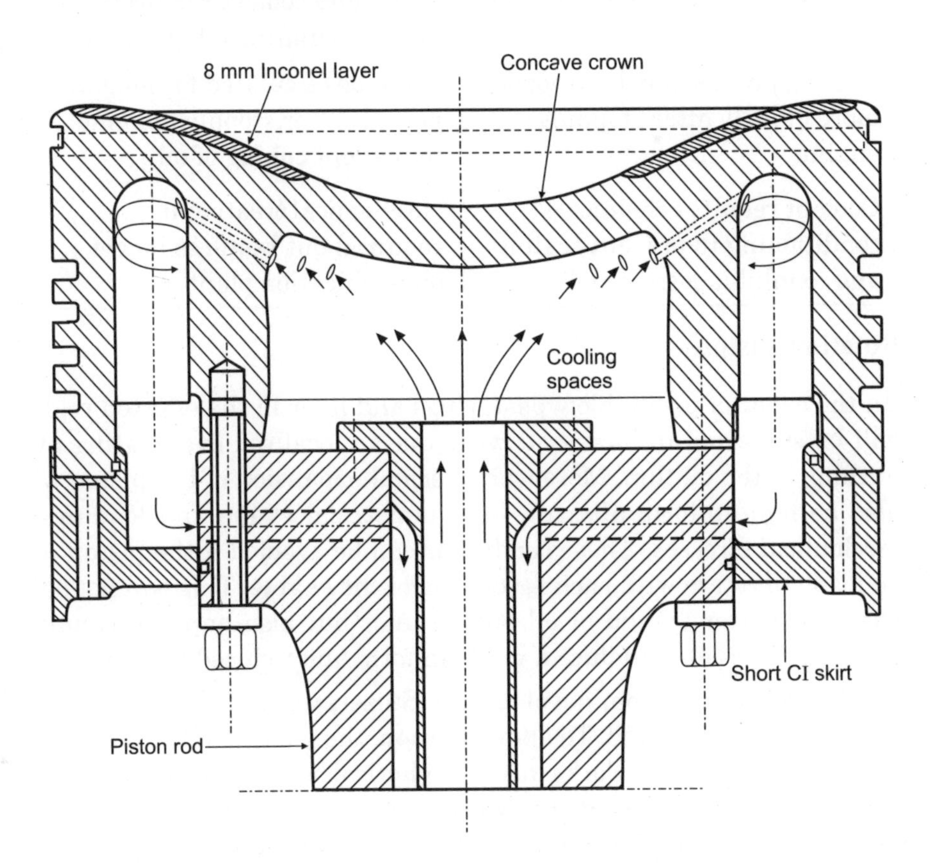

Fig. 3.9.4 Oil-cooled Piston – M A N B&W M C C engine.

Both cooling mediums have advantages and disadvantages. A comparison is given below.

Water	Oil
High specific heat capacity, removes more heat per unit volume. Outlet temperature of 70° C can be sustained	Low specific heat capacity. Maximum safe temperature of only 56° C to prevent carbonising of oil
Pistons design is more complicated and risk of water leakage into crankcase and oil contamination	Simpler design of piston because of lower thermal stresses and no risk of oil contamination
Separate piston cooling system with pumps and coolers required	Larger capacity lube oil pump and coolers required and large circulating oil quantity
Cooling pumps may be stopped more quickly after stopping engine	Oil to be circulated for longer periods after stopping engine to prevent coking of oil
Coolant requires chemical conditioning to prevent scaling and to inhibit corrosion	No chemical treatment required but requires bigger oil purification plant

Rotating Piston

In the event of piston blow past, rings and liner in the effected region will be subjected to increased wear due to localised overheating and burning of the oil film. Moreover, thermal expansion of the piston will also be non-uniform altering its symmetrical shape and leading to mechanical stresses. These conditions can be avoided if the piston can be made to rotate with each revolution of the engine. This will not only ensure a better spread of oil on the liner but also improve running characteristics and produce a more uniform and reduced liner wear.

The rotating piston, developed by Sulzer, works by making use of the swing of the con-rod. To allow rotation, the top end of the piston rod is made spherical that works inside a spherical bearing inside the piston. A pair of two spring-loaded pawls engage the teeth of a ratchet ring attached to an annular spring that is bolted to the inside wall of the piston. The force from the swing of the con-rod is absorbed by the annular ring which then transmits a compensating motion evenly to the piston. During each swing of the con-rod the pawls push the ratchet ring teeth rotating the piston by half the tooth pitch, thus creating a rotation of one tooth pitch for each engine revolution. These slow-rotating pistons (about 6 rpm) have proved very reliable in service.

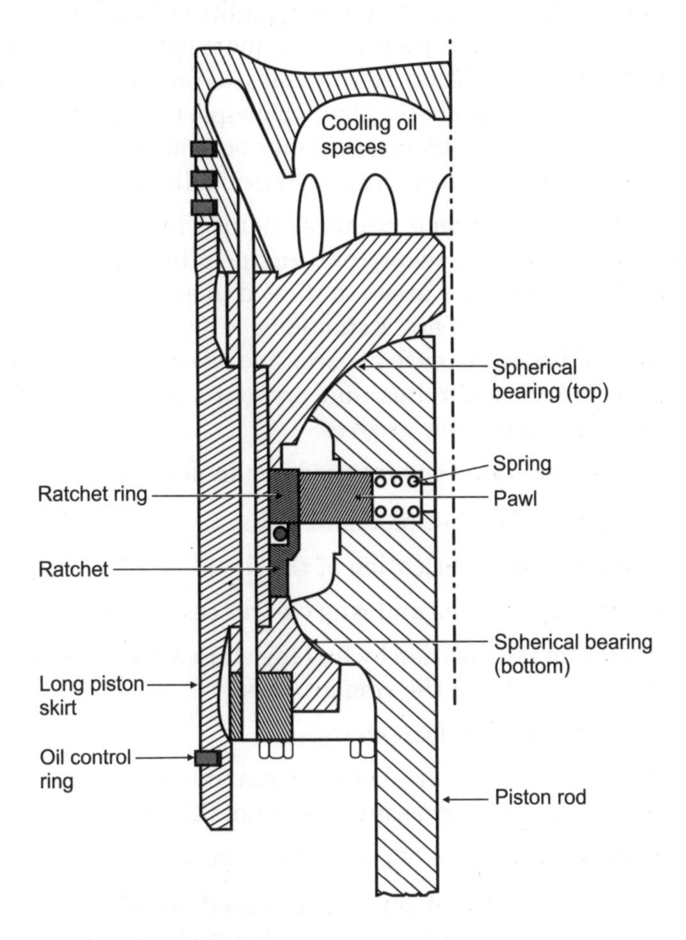

Fig. 3.9.5 Rotating piston.

Maintenance

In conventional engines, piston is withdrawn for inspection (and the liner gauged) after 4500-6500 hours for large two-stroke engines. However, this will depend on number of factors, such as, engine size, type of coolant, normal running speed of the engine, fuel oil and cylinder lube oil used and whether the engine is two-stroke or four-stroke.

Piston crown and rings need careful examination. Heavy carbon deposit on the crown or burned edges would indicate faulty combustion, poor lubrication or inadequate cooling. Rings should be free in the grooves and appear well oiled and smooth on the rubbing surfaces. If faults are observed, reasons should be ascertained and corrected. For example, if a broken ring is observed it could be due to a number of reasons, such as: worn ring grooves or liner, deposits in ring groove, combustion knock due to improper timing or poor ignition quality of fuel, excessive clearance between piston and liner, poor or inadequate cylinder lubrication etc. If on inspection the condition of the rings is found unsatisfactory they must be renewed, as otherwise piston blow-past will result which in turn can lead to a series of problems: the rings become dry as lubricating oil is blown away and due to the increased friction, liner wear will accelerate.

High-speed four-stroke engines require less frequent inspection compared to slow-speed two-stroke engines. In addition to crown and rings, there are two other areas that need to be examined. Scraper ring edges should be sharp, if not it should be replaced. Gudgeon pin should be smooth and bright. In case of excessive wear or appearance of ridges it should be replaced. Circlips that prevent axial movement of the pin should fit snugly into their grooves.

Following conditions ensure good piston performance:

- Even load distribution on all units.
- Avoiding over lubrication of the liner.
- Clean ring grooves and ring clearances within limits.
- Optimum atomisation by fuel injectors without dripping to avoid carbon deposit in the ring grooves.
- Avoiding sudden load changes.
- Differential pressure between maximum combustion pressure and compression pressure less than 32 bar.
- Sufficient quantity of clean scavenge air.

The old philosophy of pure time-based (run-hours or calendar) maintenance program is not applicable for modern ships. Classification societies require substantive proof that the machinery is in good operational condition. This does not necessarily entail opening up of the machinery. The current philosophy is to go by condition-based machinery maintenance schedule. The engine maker, of course, suggests a time chart but will always put a clause that states 'subject to operational conditions.' Hence maintenance time chart and the latest service letters issued by the engine maker are not strictly adhered to but used more for reference. Maintenance intervals, for all engine parts, may vary not only based on the design and make of the engine but also on other factors like engine condition, the type of fuel being burnt etc as mentioned above.

Stuffing Box

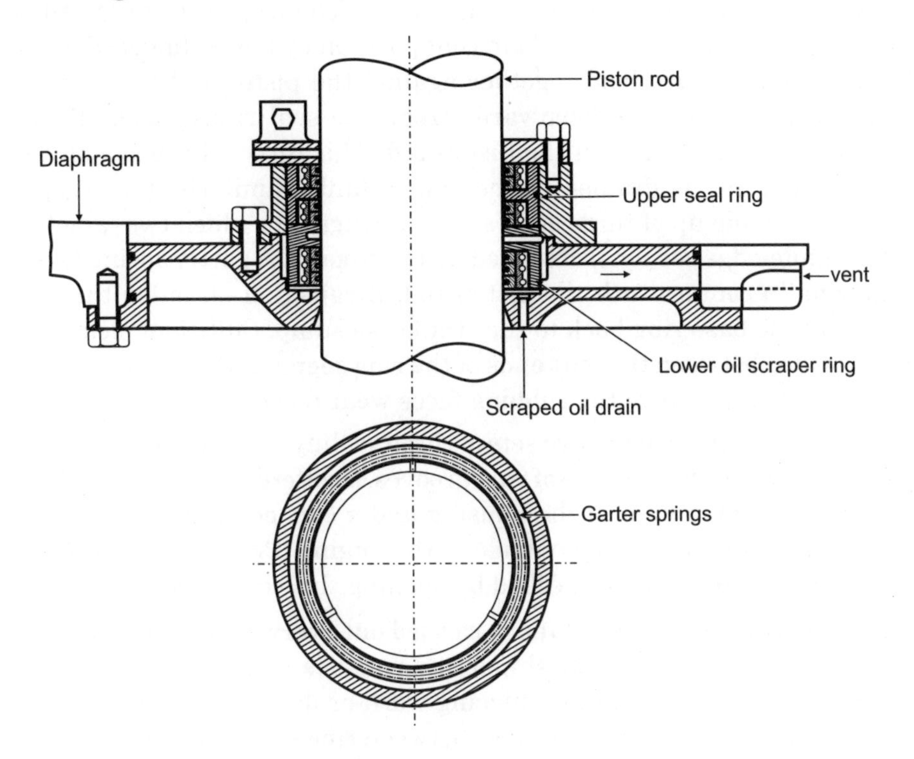

Fig. 3.9.6 Stuffing Box.

To prevent contamination of the crankcase by combustion by-products and also to prevent mixing of cylinder oil and crankcase oil it is necessary to segregate the two. A diaphragm is fitted to isolate the under-piston scavenge spaces from the crankcase. A sturdy horizontal partition extending the full length of the engine and forming the crankcase top, this may form part of the cylinder jacket in large two-stroke engines. The piston rod passes through a bore in the diaphragm where a stuffing box is mounted, Fig. 3.9.6. This works as a gland preventing lubricating oil from being drawn up from the crankcase into the scavenge air space and also preventing scavenge air from leaking into the crankcase. The cast iron stuffing box housing, which can be split vertically and is mounted on a ring bolted to the diaphragm, is in two parts – fitted respectively with air sealing rings and oil scraper rings. Five ring grooves are machined inside the housing, out of which the two uppermost ones accommodate sealing rings that prevent scavenge air from blowing down and in the three lower grooves scraper rings are fitted which scrape the lubricating oil on the piston rod as it moves upward. The sealing rings are made up of four brass or lead-bronze ring segments, which in turn accommodate eight brass sealing ring sections, two per base (some engines have teflon packing instead of brass rings). The rings are staggered in relation to each other and held in that position by four cylindrical guide pins. The parts are held together against the piston rod by a helical garter spring. On the downward stroke the seal rings crape off any residual oil and dirt from the piston rod. This dirty oil collects on the diaphragm and is drained to the sludge tank. Similarly, the scraper rings are made up of three steel segment rings into which two lamellas (thin knife-edge sections) are fitted inside grooves. A garter spring keeps the ring in contact with the piston rod. Scraped off oil is led through ports in the base ring back to the crankcase sump. Sufficient clearance is provided between the butt ends of the ring segments to ensure contact with the piston rod as the rubbing faces wear down.

Larger engines may have sets of three sealing rings and four scraper rings. Between the two sets of ring grooves, a cofferdam is machined out which, through a bore in the housing and a connecting pipe, connects with a control cock on the outside of the engine. Proper functioning of both sets of rings can be checked by opening this control cock.

Maintenance of the gland is carried out every 6000–8000 running hours (on old conventional ships) and consists of cleaning the rings, readjusting clearances and replacing worn or damaged elements. It is important to ensure butt clearance between ring segments and the axial clearance. Garter spring tension should be checked and drains cleared. During piston overhaul, the stuffing box is taken out together with the piston rod, but can also be dismantled for inspection with the piston in place. A worn stuffing box and excessive leakage can increase the risk of scavenge fires and even crankcase explosion.

Crosshead

The function of the crosshead is to convert the reciprocating motion of the piston rod into the swing of the con-rod. A pair of journal bearings that house a crosshead pin facilitate relative motion between the two, acting as a hinge to deflect the thrust of the piston rod via the con-rod to rotate the crankshaft. To protect the piston rod from the lateral thrust produced by the swinging con-rod a pair of guide shoes or slippers are mounted on stepped journals either end of the pin that transmit the thrust onto vertical guide plates attached to the engine frame. The slippers float on the journals, allowing for slight misalignments in the guides. Made of cast steel and lined with white metal on the sliding surfaces, the slippers keep the piston perfectly aligned as it reciprocates inside the liner.

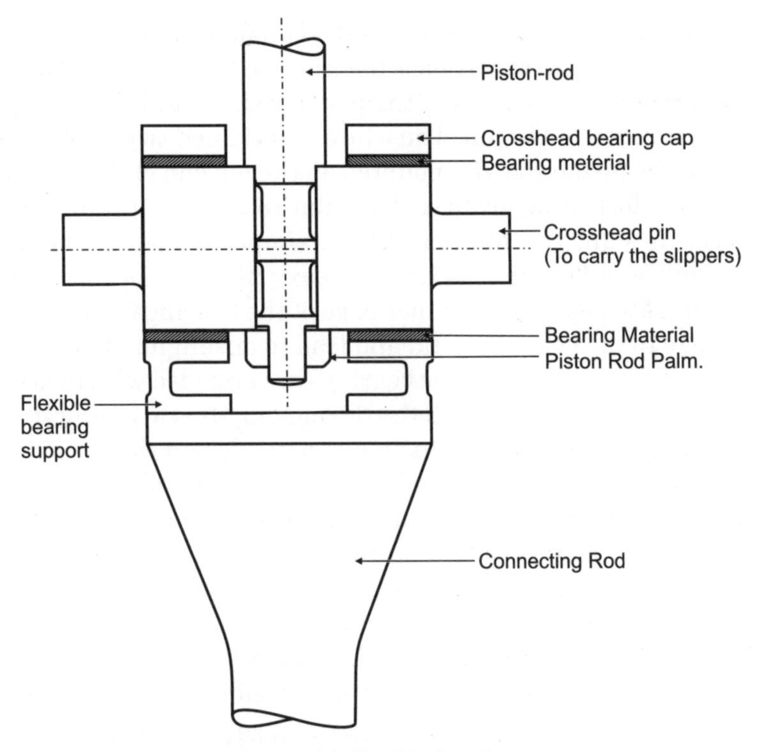

Fig. 3.9.7 (a) Crosshead with flexible bearing support.

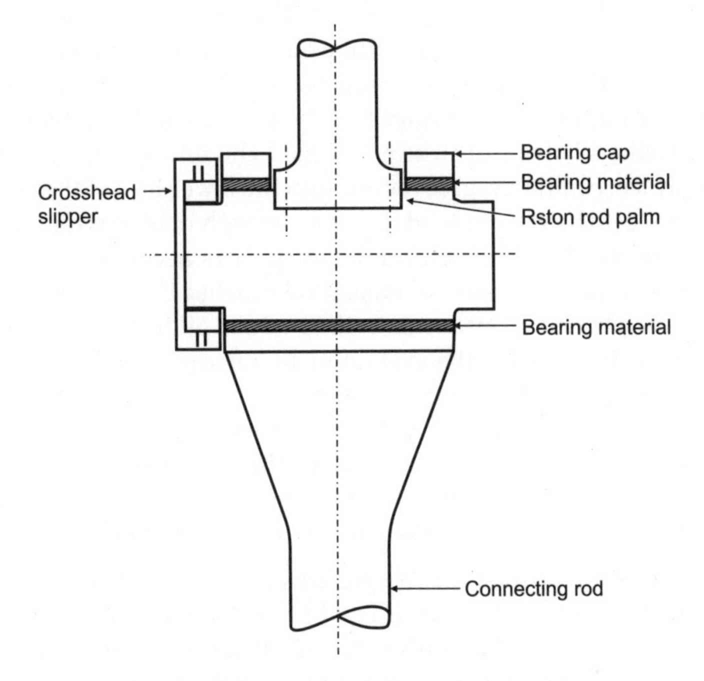

(b) Crosshead with one-piece lower bearing

The crosshead pin is of forged steel ground and finished on the journal portions. The central portion is flattened on four sides and bored to receive the threaded bottom end of the piston rod. A nut below the pin secures the piston rod to the pin. The white-metal lined twin bearings on either side of the piston rod are mounted on the top end of the con- rod – which is either forked or given a flat palm to support the bearings. In the forked design, the pin acts like a simply supported beam with the possibility of bending under the heavy piston load in the centre. The concentrated load on the inner edge of the bearings could cause fatigue, damage to the bearing metal and lead to misalignment and unbalanced forces. To overcome this, the bearings are provided with flexible supports allowing them to align with the flexing pin, thus avoiding stress at their inner edges (Fig. 3.9.7 a). In an alternate design the top end of the con-rod itself forms the bearing housing – a single-piece bottom half over the entire length of the crosshead pin. The bottom of piston rod is shaped like a foot which is bolted on to a niche cut on the top of the pin in between the bearings. A pair of bearing caps, bolted to the bottom housing, covers the top halves of the pin The large diameter pin or the journal, which is fully supported at the bottom and with the increased effective bearing area, eliminates any possibility of bending (Fig. 3.9.7 b).

Since the thrust is almost continuously downward, the oil film on the bottom half of the bearing will tend to squeeze and thin out. To compensate, axial oil gutters are cut into the shell of the bottom half. Oil from the medium-pressure engine lube oil system is supplied to the sides of the crosshead pin (which is hollow bored) through articulated pipes. Oil flows through drilled passages to the two journal bearings and the slippers. Drilled holes in the slippers lead the oil to grooves cut on its sliding surface. Similarly, grooves cut into the lower half of the crosshead bearings collect the oil and lead it down through drilled passages in the con-rod to lubricate the crank-pin bearing. It is necessary to maintain sufficient oil supply pressure to ensure oil reaches the various bearings. This is accomplished by either installing a lube oil booster pump taking suction from the engine lube system or by incorporating an additional plunger-type crosshead oil pump mounted near the top end of the con-rod and actuated by its swing. The angular movement between the con-rod and the crosshead drives the pump through a linkage. The high-pressure pulse of oil delivered to the crosshead bearings also tends to keep the journal 'floating' all the time extending its life.

Four-stroke trunk-piston engines have a gudgeon pin to carry out the function of the crosshead. The hardened steel pin has a polished bearing surface. It may be either free-floating or fixed in the piston-skirt and works inside a bearing sleeve pushed into the eye of the connecting rod at its top end. The bearing is lubricated by oil travelling through a drilled passage in the connecting rod from the bottom-end bearing which in turn communicates with oil holes in the crankshaft.

Maintenance

Maintenance of the crosshead involves periodic inspection of the journal bearings. In a process called **wiping**, part of the white metal contact faces may get wiped out, the wiped out material getting lodged in the oil grooves. Corrective bedding of the lining material to increase area of contact between journal and lining and to ensure that the weight is evenly spread across the bearing surface is the solution. If the bearings are of the thin-walled shell type and if the insert is badly worn or cracked due to poor bonding of white metal to steel, the insert should be replaced. Clearance between the slippers and guide ways should also be checked using feeler gauge since an increase in this clearance or any misalignment in the guides could cause excessive wear between piston rod and stuffing box and also between piston and liner. Maximum allowable clearance between guide and slippers is 0.7 mm. Shims can be inserted between the guides and the engine frame to bring the guides forward and reduce the clearance.

Connecting Rod and Crank Pin Bearing

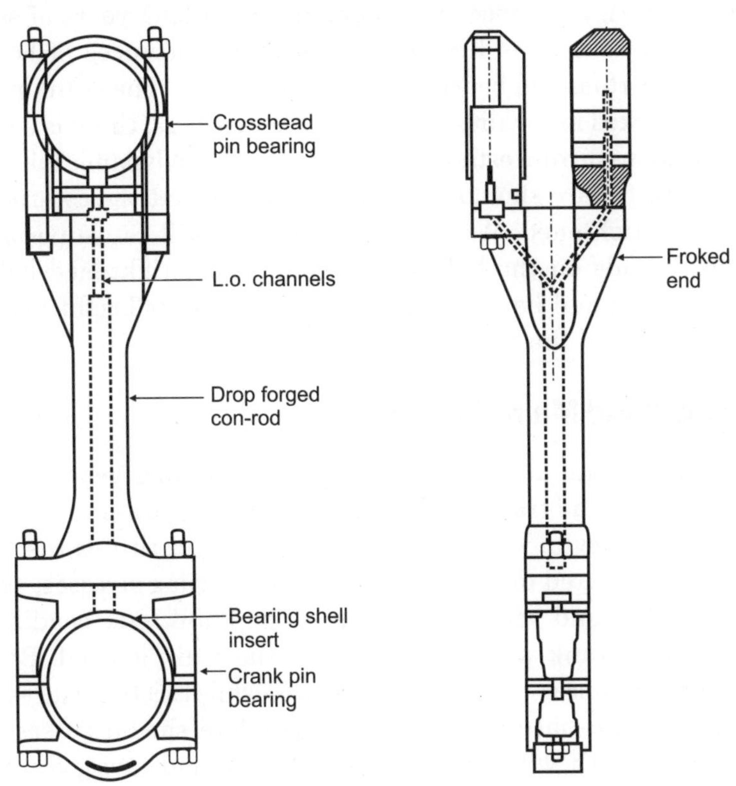

Fig. 3.9.8 Connecting rod.

The connecting rod connects the crosshead with the crank pin, transmitting the combustion force to the crank to rotate the crankshaft. In older designs the con-rod had a flat palm at both its ends; the crosshead or the top-end bearing housings bolted to its top end and the crankpin or bottom-end bearing housing bolted to the bottom end, as shown in Fig. 3.9.8. Later designs had a single forging, incorporating a single-piece bottom half of the crosshead bearing housing and top half of the crank pin bearing housing. Again, older engines carried white-metal lined thick-walled bearings scraped to fit, where clearances were adjusted by adding or removing shims in between the bearing halves. Later designs had thin shells; lined with a 1 mm thick layer of white metal, held between the bearing housing and a steel cover (a thin layer of white metal bonded to steel affords much increased fatigue strength and load capacity). When clearances increased, the shells were simply replaced. Lube oil from the crosshead bearings flowed down through a drilled passage in the con-rod to reach the crankpin bearing.

Maintenance is hardly required for the con-rod itself, since it is a single-piece drop-forged component. However, during overhauls, landing faces of bolt heads should be closely examined for creeping cracks (brought on by stress fatigue), especially on engines with long years of service. Regular maintenance is limited to checking clearances of both the top-end and bottom-end bearings every six months. The crank pin also should be checked for ovality, since if this is in excess, there is a danger of the continuous oil film established due to the hydrodynamic lubrication breaking down. Care should be taken to ensure that the con-rod bolts are correctly tightened. While over tightening will cause permanent injury to threads on the bolt and nut, an under-tightened bolt will produce rubbing and hammering which will show as bright marks on the bolt shank and the head and nut.

Crankshaft and Main Bearings

A crankshaft is the mainstay of the engine running gear, weighing hundreds of tons and transmitting the full working power of the engine generated in the cylinders to the flywheel that constitutes the engine output. It is subjected to a combination of fluctuating stresses, such as bending, torsion and shear – bending of the crank pin, bending and twisting of the crank web and torsion on the main journal. Bending causes tensile, compressive and shear stresses while twisting causes shear stresses. The material it is made from therefore should possess high mechanical strength, withstand prolonged fatigue and have good bearing surfaces. Depending on the size, either low carbon (0.2 % to 0.4 % carbon content) or low alloy steels (chrome-molybdenum and nickel) are generally used.

A crankshaft is made up of number of cranks or 'throws', rotated by the con-rod, one crank for each cylinder, for in-line engines (V-engines can have two con-rods per crank pin). Every crank has two webs, with the crank pin held in between at one end and on to which the crank pin bearing is attached. At the other end, the web is shrunk-fitted onto a journal, to form part of a continuous shaft. Each of these journals is supported by a main bearing located in the transverse girder of the bedplate. The shaft is thus made up of forged and machined main journals and crank webs, to form individual cranks. These are manufactured in one of three ways: 1) semi-built up, where a forged one-piece crank (consisting of two webs and a crankpin) is shrunk fitted onto journals (cold journals inserted into bored webs heated to 400° C); 2) fully built-up, where individual webs are shrunk fitting on to both journals and crank pins; and 3) welded, where webs, journals and crankpins are all welded together, as illustrated in Fig. 3.9.9. The last method is most suitable for large two-stroke engines because it offers increased stiffness of the crankshaft to carry the much heavier torques. These shafts also will have a higher natural frequency of torsional vibration.

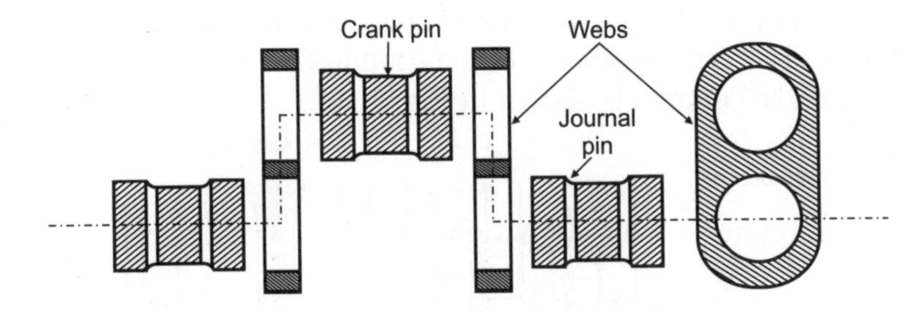

Fig. 3.9.9 (a) Fully built-up crankshaft.

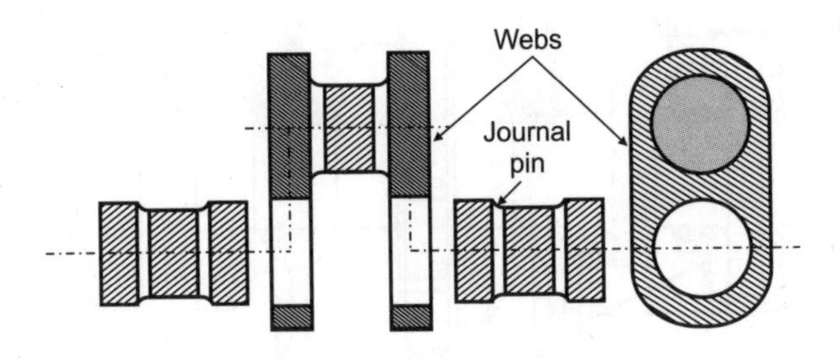

(b) Semi built-up crankshaft

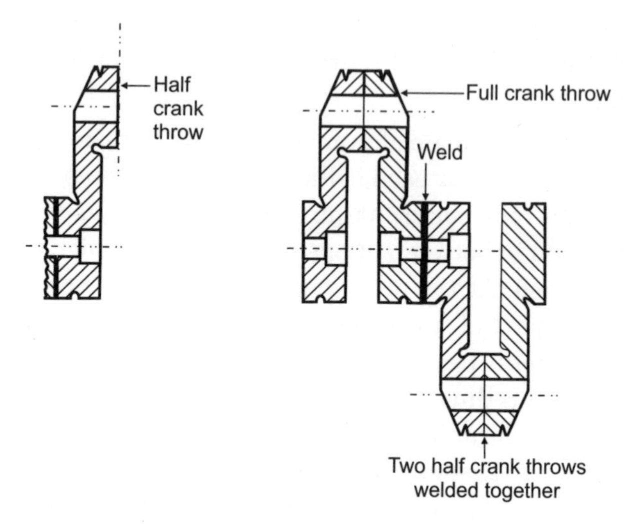

Fig. 3.9.9 (c) Welded crankshaft.

Various jigs ensure correct crank angle during assembly. External balance weights may be attached to the crank webs (or holes bored through crank pins) for dynamic balancing of the crankshaft. Fillet radii are machined at each end of pins and journals to eliminate stress concentration points that could lead to fatigue cracking. The finished crankshaft is subjected to thorough tests using ultra-sound and other techniques. One end of the shaft carries the flywheel while the other end is free. Gear or chain-drive sprocket wheels are mounted on the crankshaft to drive the camshaft. The longitudinal section of a crankshaft of the Sulzer RND engine is shown in Fig. 3.10.

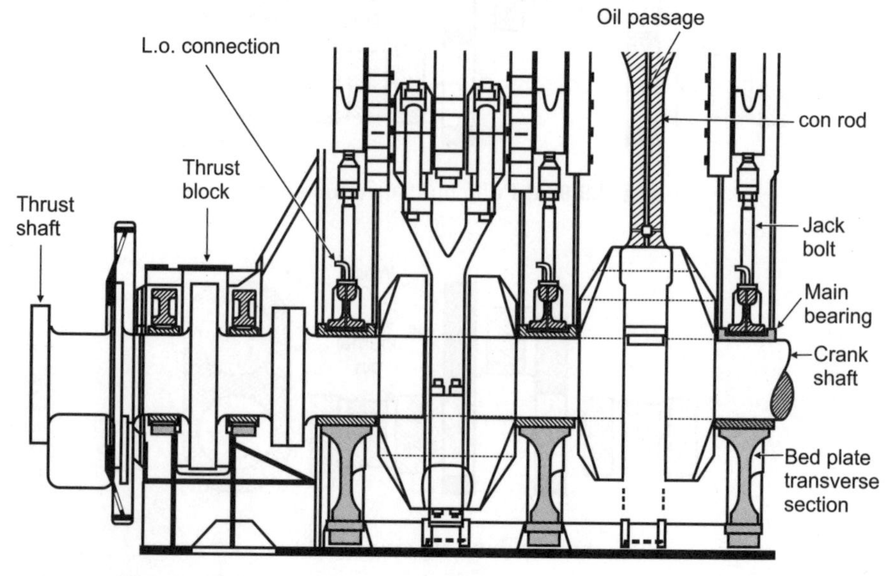

Fig. 3.10 Crankshaft and main bearing of Sulzer RND engine.

Four-stroke medium and high-speed engines generally have one-piece cast or forged crankshaft. Solid forging by the continuous grain process, where a single billet of steel is forged into the finished shaft, offers improved fatigue resistance.

Maintenance

In shrink fitted crankshafts reference marks are provided on the crank pin and webs to show their relative position. These marks are checked during crankcase inspections to find if any slippage has occurred. Slippage can occur in the event of an obstruction to the rotation of the crankshaft, brought on by fouling of the propeller, sudden stoppage of the engine or blowing through the engine with the turning gear engaged. Slippage will effect the engine timing and if not corrected will result in inefficient operation and poor starting.

The main bearings (and the bottom-end bearings) are checked at definite intervals as specified by the classification societies. In large two-stroke engines a retractable feeler gauge is inserted between the crankshaft journal and the bearing top-half to check the clearance. Clearance should be 0.5–0.6 / 1000 of the bearing diameter. However, for an engine with a crankshaft diameter of 990 mm, a clearance between 0.45 – 0.75 mm is sometimes maintained. Additionally, a depth gauge is used to check the drop of the crankshaft journal due to wearing down of the bottom-half of the main bearing. An increase in wear would indicate faulty lubrication, poor quality lube oil or contamination of the lube oil with fuel oil (10% is considered the allowable maximum). Clearances are set using shims inserted between bearing shells and the housing. Next, the alignment of the crankshaft is checked by measurement of the crank deflection for each throw in four positions. Care must be taken to ensure that the journals are resting on the main bearings when the readings are taken. The readings are compared with the original deflections recorded in the engine manual. Misalignment could occur for a number of reasons, such as: worn main bearings or incorrect clearance adjustment, distortion of engine bedplate or damage to ship structure. Any of the above faults – misalignment of the crankshaft, excessive main bearing clearances or slipped crank web on journals – will directly result in increased engine vibration, and can lead to high stresses and result in fatigue failures.

A careful examination is to be made of all crankpins and journals for cracks especially at the fillets and around the oil holes.

Main bearings support the crankshaft at each journal and are therefore subjected to heavy fluctuating loads that are transmitted through the crankshaft from adjacent units. They are supported on transverse girders in the bedplate, rigid enough to withstand both vertical and transverse loads. The bearings are generally of the thick-walled white metal type, the top half of which is either bolted down on to the bedplate by studs and nuts or held by jack bolts tightened in between engine frame and the bearing cover. Four-stroke medium speed engines and modern two-stroke engines use thin-walled steel-backed bearings. These bearings are changed when the clearance reaches a maximum.

Maintenance involves checking bearing clearances using a feeler gauge and checking the crankshaft drop using a depth gauge. Excessive clearance or a badly pitted white metal lining will necessitate replacement of the bearing shells (clearance can be adjusted using shims placed between the shell and the housing). While the top half of the bearing shell is easy to remove, the crankshaft will need to be lifted by about 0.1 to 0.15 mm using hydraulic jacks in order to turn out the bottom half. This will avoid scoring the back of the shell and the bearing housing. To jack up the crankshaft, a transverse beam is placed below the crank-web. Two jacks are then supported on it placed either side of the bottom-end bearing against the crank-web next to adjacent main bearings, both connected to the same hydraulic unit. Care should be taken to ensure that the jacks are in line with the crankshaft centre.

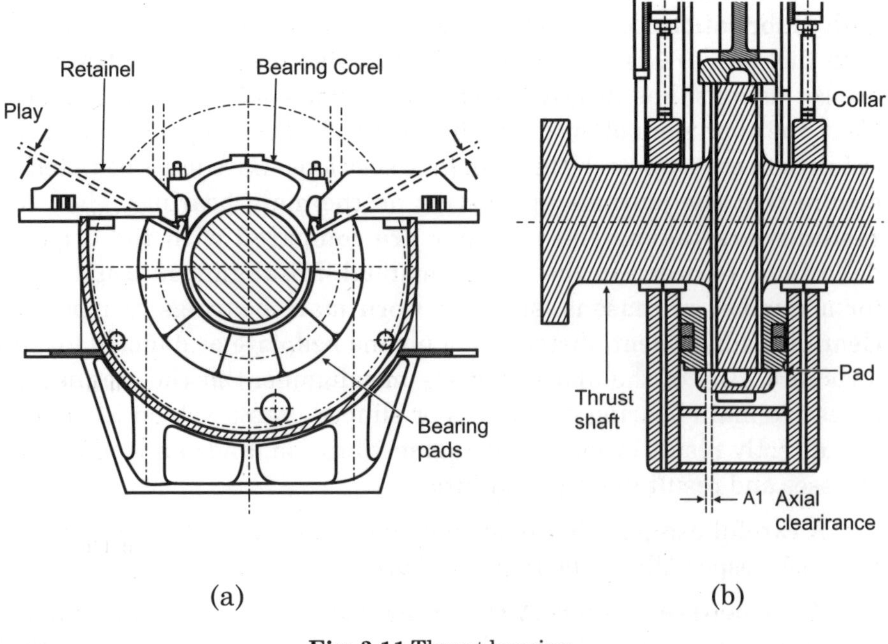

Fig. 3.11 Thrust bearing.

Thrust Bearing

Thrust bearing is provided in an engine that is directly coupled to the propeller in order to safeguard the engine from the transverse thrust of the propeller, generated as an equal and opposite reaction to the thrust imparted on the water. It is accommodated either within the engine or at the aft end of the crankshaft outside the engine frame and transmits the propeller thrust safely to the engine structure to which it is bolted. The bearing is usually a single-collar tilting pad type carried in a housing that is integral with the aft part of the bedplate. The large diameter collar on the aft end of the thrust shaft that is bolted to the propeller shaft forms the thrust collar, (Fig. 3.11 (b)). The collar is solid forged with the thrust shaft which is supported on two journal bearings. White metal lined thrust pads held in recesses in retainer segments inside the housing (accurately machined and perpendicular to the shaft) bear against the collar and transmit the thrust from the collar to the housing. The collar should have an axial clearance of 0.5 mm on either side which is adjustable using shims behind the ring on which the pads pivot. From the housing the thrust is transmitted to the bedplate and on to the ship's structure. The thrust bearing is lubricated from the engine lube oil system, the collar running in an oil bath. Oil from the pads drain to a pan at the bottom from where it is led through a pipe to the engine crankcase. Thermometers in the housing help monitor temperatures and trigger alarms in case of over heating.

Thrust bearing clearances are checked periodically to ensure they are within limits. The engine is first blown through in the 'Ahead' direction. This pushes the crankshaft forward due to the counter thrust exerted by water on the propeller. This will make the thrust collar butt against the forward thrust pads. The total displacement of the collar is measured using a clock gauge. Next, the engine is blown through in the 'Astern'. direction so that the thrust collar butts against the aft thrust pads. This displacement of the collar is also measured. Both the readings will be between 0.8 to 1.3 mm for a new engine. This allows for expansion of the crankshaft and helps in maintaining an optimum clearance between crank webs and adjacent main bearings. Maximum allowable thrust bearing clearance due to wear is 2.5 mm.

Flywheel

Flywheel is fitted to the end of the crankshaft outside the engine frame. In medium and high-speed engines its function is to store energy while the engine is running, the stored energy being proportional to the square of the engine speed. The high mass and inertia of the flywheel stores the energy during peak torque and returns it to the engine during low torque to push the piston up and compress the charge air in the cylinder. Thus, by dampening the non-uniform torque caused by the individual cylinders firing, it also smoothens the fluctuations in the engine speed. In large slow-speed engines however, the inertia of the rotating parts of the engine itself contributes much of the energy required to push up the piston during compression stroke. In these engines the flywheel is provided with gear teeth on its rim that engage with the turning gear to turn the engine over during overhauls.

Camshaft

Camshaft is almost always located at the engine mid-platform level, though the position dependents on the engine size and type. It is driven from the crankshaft, accurately synchronised through either a train of spur gears or a chain drive located at the aft end for engines with up to six cylinders and at the mid section for others, as shown in Fig. 3.12 (a) and (b). Two stroke engine camshafts rotate at the engine speed while those of the four-stroke engine rotate at half the engine speed since one complete engine cycle is completed in two engine revolutions. Also, two-stroke engines have lesser number of cams since there are no air inlet valves whereas four-stroke engines have more cams. Large two-stroke engines use the chain drive while most medium speed engines use the gear train. The bearings of the intermediate gears are lubricated from the engine low-pressure lube oil system and a spray nozzle lubricates the gear teeth. The gear on the crankshaft is in halves, while the intermediate gears (there would be two of these in larger engines) are in one piece. The bearings of these gear wheels and the gear teeth are lubricated from the engine L.O. system. In case of a chain drive, the transmission system comprises of two matched roller chains of identical size and a tensioning device fitted to the chains to preserve the required tension. A spray nozzle injects lube oil between the chain sprocket wheels and the chain rollers, before the rollers engage the wheel to provide an oil cushion. The starting air distributor, the governor, cylinder lubricators and indicator gear are driven from the chain drive intermediate shaft, through further chains and gears. In the case of gear drive, a vertical shaft is taken from the camshaft through a pair of helical gears for this purpose.

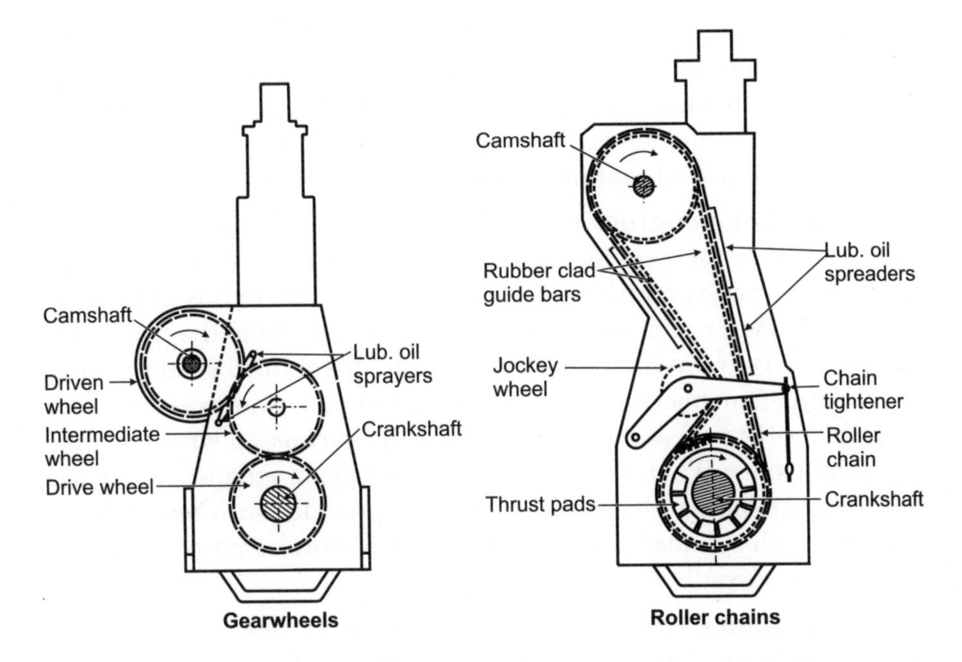

Fig. 3.12 Camshaft drives. (a) Gear drive (Sulzer). (b) Chain drive (B &W).

Reversing of the engine is performed in one of two ways. 1) By axially shifting the camshaft through a reversing piston incorporated at the free end of the camshaft and actuated by hydraulic pressure. The camshaft in this case carries a set of 'Ahead' and 'Astern' fuel and starting cams for each cylinder (M A N engines). As the camshaft shifts, a new set of cams slide into position under the followers, without the follower rollers being lifted. 2) A single cam is used for both 'Ahead' and 'Astern' directions and the correct timing is achieved with the use of an oil-operated reversing servomotor, working as a lost-motion-clutch, incorporated in the camshaft gear wheel. In this case, the clutch rotates the camshaft by a certain degree relative to the crankshaft to bring the cams in position for reverse firing of the cylinders (Sulzer engines).

Hardened and ground cams mounted on the camshaft operate the fuel pump plungers and other valves that control the engine cycle. Cam profile, shape, angle, base circle, dwell etc are designed according to the required valve timings, follower lift and acceleration and deceleration at the beginning and end of strokes. The profiles are such that the transition between the various stages in a stroke is smooth without any bouncing or shock. Each cam is in halves and is mounted on a sleeve keyed to the shaft. Radial serrations on one side of the cam engage with similar serrations on a flange on the sleeve and the two are bolted together to secure the cam on the shaft. The angular position of the cams can be adjusted when required. The camshaft is sometimes made up of individual sections, one for each cylinder, all flanged together and supported rigidly in force lubricated white-metalled bearings, split vertically to enable easy removal. Some engines have a separate camshaft lube oil system to prevent crankcase oil contamination from fuel oil leaking from pumps.

Four-stroke engines have additional cams for inlet and exhaust valves and they rotate at half the speed of the crankshaft since a cycle is completed in two revolutions of the engine. The camshaft is ground to a uniform diameter over its entire length and driven from the crankshaft by cast iron gearwheels supported by white-metal lined steel bearings. Three hardened steel cams, one each for the inlet and exhaust valves and one for the fuel pump, are hydraulically shrunk-on with the aid of tapered bushes. When used as a main propulsion engine in a geared multi-engine configuration, the engines are provided with sets of double cams for each cylinder along with a hydraulic reversing gear to move the camshaft axially for reversing.

Maintenance consists of checking for wear, bearing clearances, wheel alignment, chain tension and seized chain rollers. While excessive chain tension will cause high loads and increased wear, a slack chain can cause vibration and possible fatigue failure. Chain elongation is therefore checked periodically and if it exceeds 2% of the original length, the chain is renewed.

Governor

A governor is fitted to an engine to control the engine speed so as to keep it at a set level, with only a transient variation between close limits while power output changes to meet demand. This is achieved by the governor by automatically changing the engine fuel pump setting to match the load, thus maintaining engine speed. Such a governor is termed **isochronous**. It also serves to prevent racing, coming into operation when the engine exceeds a predetermined speed. Earlier, only diesel generators were fitted with simple flyweight-controlled governors and large two-stroke engines, because they were inherently stable, were provided only with over speed trips that shut down the engine in the event of over speeding (about 15 % above normal rpm) due to a sudden shedding of load in an emergency such as propeller shaft fracture or propeller emerging from water in rough weather. Modern engines, both low-speed two-strokes and high-speed four-strokes have sensitive hydraulic governors of either the Woodward type or electronic. In the case of main propulsion engines, where the set speed can be adjusted, they are termed **variable speed governors**.

A governor essentially has to perform two separate functions: act as a speed measuring device and, regulate the fuel in proportion to the changing load to automatically control the speed. The first is accomplished by a pair of spring loaded flyweights that is rotated by a drive from the camshaft and the second, by a servo-motor oil pressure that transmits the regulating force, triggered by the movement of the flyweights, to the fuel pump regulating shaft through a system of links and rods.

Proportional Action Governor

A purely mechanical flyweight type governor will be incapable of maintaining the set speed under varying load, because of the inertia of its components, friction between them and the non-uniform spring compression load acting on the governor spindle. There will be variation in speed following changing load conditions. The difference between the set speed and the new equilibrium speeds obtained after every change in fuel pump setting triggered by the governor is termed **offset**. And the drop in speed from the no load to the full load conditions is called the **speed droop.** A power amplifying servo-system is therefore incorporated to overcome the variation in speed and reduce the offset. The control action in these governors is obtained by the **closed loop control system** (Fig. 3.13), wherein the engine speed is compared with the set value and if there is a deviation the governor produces an output to increase or decrease the fuel pump output proportional to the deviation, in order to keep the engine speed close to the set value. This is called **proportional control**.

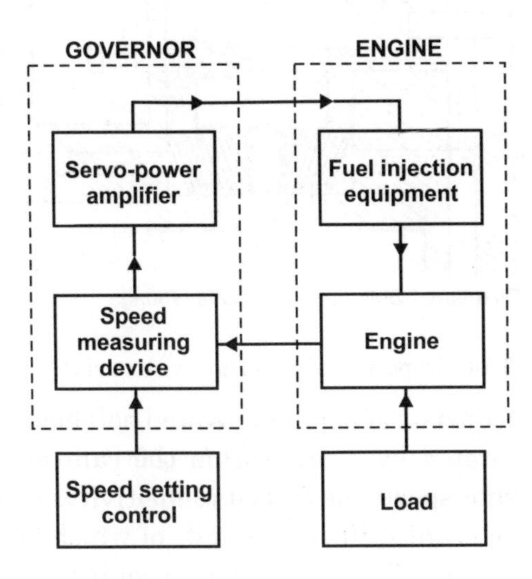

Fig. 3.13 Basic arrangement of the proportional control system.

As shown in the block diagram 3.43, in the event of a momentary speed drop due to sudden increase in load, the speed measuring device will obtain a corresponding signal from the engine which it will then compare with the value set on the speed setting control and the deviation found will produce a corresponding action from the servomotor-controlled power amplifier of the governor which will then reposition the fuel tack, thus increasing fuel supply to meet the increase in load.

The servomotor, through which the governor actuates the fuel pump control links, amplifies the movement of the flyweights making the system very responsive to changing load conditions. The hydraulic servomotor system not only reduces the time between a load change and fuel alteration, but also positions the fuel rack precisely to push the engine speed toward the set value.

A schematic diagram of a servo-controlled isochronous governor is shown in Fig. 3.14.

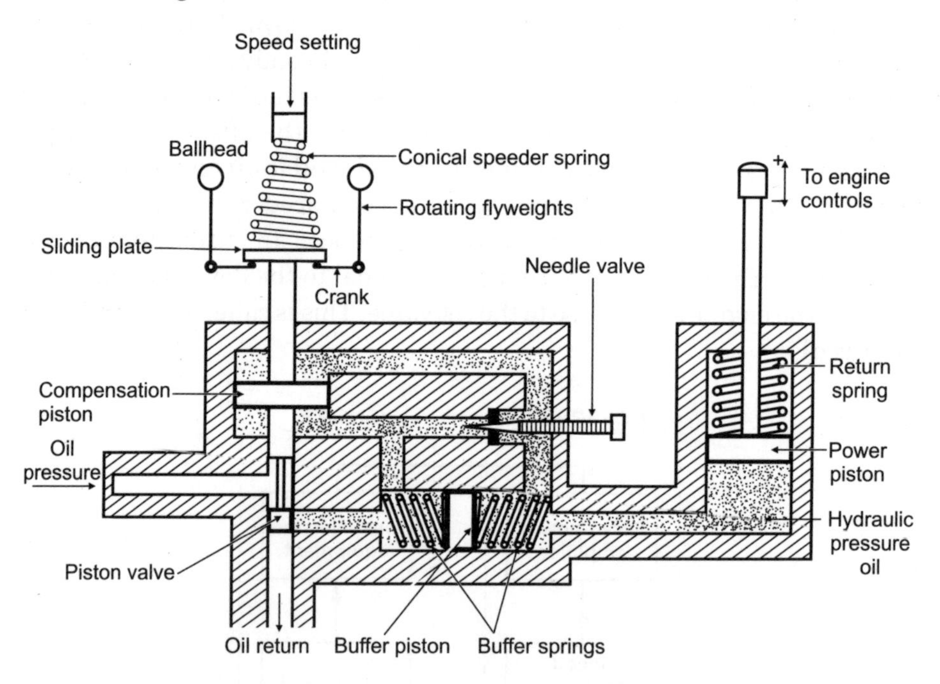

Fig. 3.14 Proportional governor with variable speed.

The flyweight consists of a pair of pivoted ball heads on either side of a sleeve that is rotated by a drive from the camshaft, stepped up to increase the governor speed and thus its sensitivity. The ball heads form one end of a pair of cranks, the other ends of which bear against a disc attached to a pilot valve. The disc is held in equilibrium under the force exerted by the flyweights and a conical speeder spring bearing down on it. The conical shape imparts a linear relationship to the compressive force of the spring and the centrifugal force of the ball heads, resulting in a linear motion of the valve spindle proportionate to the changing engine speed.

An increase in speed resulting in increased centrifugal force on the ball heads will draw them apart, and the cranks will push the disc up. And a drop in speed will draw the ball heads nearer, lowering the cranks and moving the disc down under the spring load. The valve spindle also carries a compensation piston. Any difference of oil pressure on its either side will set up a force due to the imbalance and push the piston up or down, adding to or subtracting from the compressive force of the speeder spring. This results in a temporary speed droop.

As the engine slows, the ball heads cause the valve spindle to move down uncovering the port and admitting pressurised control oil into the servo-system. Oil pressure acts on the buffer piston pushing it inside the buffer cylinder, compressing the right side buffer springs and forcing oil into the power piston. This raises the piston which in turn actuates the fuel pump control rod to increase the engine speed. Compression of the buffer spring on the right side of the buffer piston makes the oil pressure on the right side slightly less than on the left. This pressure difference acts directly on either side of the compensation piston and the imbalance pushes the piston up, against the speeder spring load. The upward movement of the valve spindle tends to close the pilot valve even before the engine reaches its set value. Next, the buffer piston regains its equilibrium position as the opposing forces of the two buffer springs even out. Similarly, as the oil bleeding through the needle valve equalises the pressures on either side of the compensation piston, this too regains its earlier position of balance. The ball heads also return to their equilibrium position corresponding to the set speed, after having pushed the fuel pump control lever to a higher setting to deliver an increased quantity of fuel commensurate with the increased load.

An increase in engine speed on the other hand, will cause the cranks to raise the valve spindle, raising the pilot valve and uncovering the port to drain pressurised oil from the servo-system. The drop in pressure on the left side of the buffer piston will make it move left and cause the power piston to move downward. This in turn will actuate the fuel pump control lever to reduce the fuel setting. Additionally, pressure below the compensation piston will drop moving it downward too, which in turn will make the pilot valve block the port. These movements are again temporary till equilibrium is established once again in the system as before. The needle valve position is adjustable allowing adjustment of the oil flow rate across it, which in turn determines the governor's response characteristics – how soon the engine responds to a change in load.

Control oil is tapped either from the engine lube oil system or from a separate gear pump operated by the governor drive from the camshaft. Where the governor is lube oil powered, any drop in oil pressure will cause the power piston to move downward under the spring load shutting off fuel supply to the engine – thus providing an additional fail-safe mechanism.

Proportional Governor with Reset Action

A purely proportional governor will still produce slight variation in speed following changing loads (to a small extent also due to the non-uniform torque produced on account of individual cylinder firing). Hence, to keep the speed constant across changing load conditions, that is, to bring the offset completely to zero, the so-called **reset action** is incorporated in the governor. Reset action, also called integral action ensures that the output signal (movement of the power piston) produced by the governor is linear, or directly proportional to the input signal (a change in the rotational speed of the ball heads). Additionally, identification and correction of speed variations due to cylinder firing differences (torsional oscillations) is also accomplished.

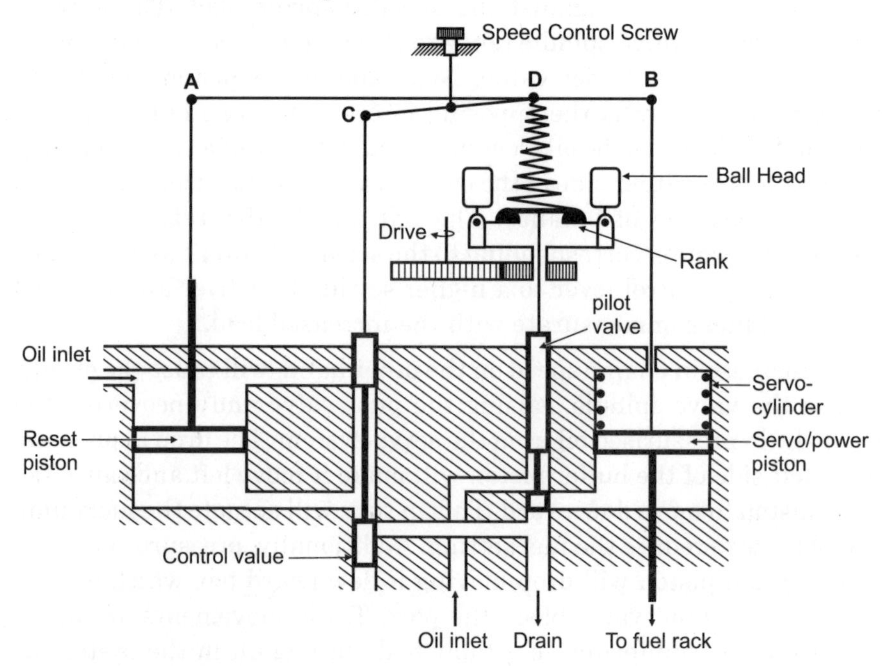

Fig. 3.15 Proportional governor with reset action.

Fig. 3.15 shows a working diagram of a governor with proportional and reset action. The compensation piston is separated from the main control valve spindle and has its own separate control valve. Buffer piston and needle valve are absent, instead, a system of linkages connecting the power piston, compensation piston and its control valve restore the system equilibrium after a speed correction.

A drop in engine speed due to an increase in the load will draw the flyweights inward opening the pilot valve and admitting pressure oil into the servo-cylinder. Following this, the power piston will move up against the spring, setting off the following events: 1. the fuel rack will shift to a new position to increase fuel supply, 2. end B of the link AB will get pushed upward, 3. end D of link the CD will be pushed up rotating it about the pivot below the speed control screw, and causing the opposite end C to move down, lowering the reset piston control valve. This will release some oil from under the reset piston into the drain. In turn, the piston will move down and as end A of link AB comes down, it will press end D of link CD downward causing the opposite end C to move up and the attached control valve to close off the drain. Thus, point D and point A will return to their original positions. The engine will now be running at its original speed but with a new, increased fuel pump setting. The speed droop during the load change/speed restoration process would be temporary. This type of governol with both proportional control and reset action is called an Isochronous governor.

Load Sensor

This is a device that can sense a change in the load on an engine and supply a proportionate signal to the governor to change the fuel pump setting even before the load change brings about a change in the speed. Obviously, the device works faster than the governor can sense a change in speed. Because mechanical devices to sense a change in the load will be too complicated and far too costly, load sensors are always electronic devices. A load-sensing element is incorporated in a diesel generator governor which can detect changes in the current flow. An increase in the electrical load will increase the flow of current and this is fed as a signal to the governor to supply an increased fuel through electro-mechanical means before the engine begins to slow, providing fast response to load changes. The speed-sensing element in this case would make corrections necessitated by small errors in rack positioning.

Electric Governor

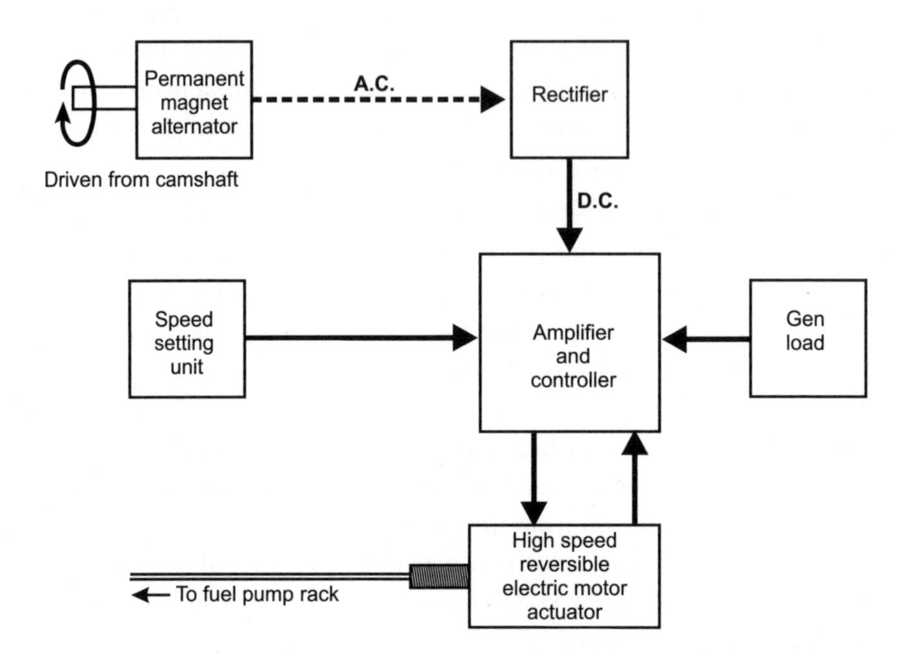

Fig. 3.16 Schematic diagram of electric governor.

The working of an electric governor is shown schematically in Fig. 3.16. This employs proportional control with reset action along with load sensing. The engine speed is picked up either by a permanent magnet alternator driven from the engine or a speed sensor such as a magnetic pickup. An alternator would generate an AC voltage proportionate to the engine speed, which is converted into DC by a rectifier, then fed into a power amplifier-cum-controller. A speed-setting unit generates another DC voltage of opposite polarity representing the set or desired operating speed, which is also fed into the amplifier. If the two voltages are equal and opposite they cancel out. If they are unequal, then a signal proportionate to the extent of difference, or error, is generated which is then transmitted by the controller to either a fully electric or electro-hydraulic actuator, which repositions the fuel pump rack. A fully electric actuator will have a permanent magnetized DC motor with built-in tacho-generator as the drive source. Instead of an AC alternator, if the speed sensor is a magnetic pickup, the signal is similarly fed into the amplifier and compared with the set speed value.

In case of a diesel generator, the input from a load sensor is also fed into the amplifier, which on detecting a change in the load condition will generate a signal that is similarly transmitted to the actuator for fuel pump resetting. All analogue input signals are monitored to be within a pre-defined range. If the signal is out of range, an alarm is simultaneously initiated and indicated on a LCD display unit. Electric governors can be incorporated as part of an overall electronic control system. Such an integrated system performs number of actions such as, over speed monitoring and suppression, critical rpm blocking and alarm, synchronizing, load-sharing etc. When adapted to a main propulsion twin-engine configuration, the system can synchronize both engines connected to a common gearbox and in case of twin propellers, it can synchronize both the propellers. When incorporated in diesel generator sets, the system synchronizes generators and facilitates load sharing, compensating for major load changes and producing only minor changes in frequency. In case there is any malfunction in the system, load sharing is accomplished by modification of the speed settings.

Digital Governor System like the DEGO III speed governor manufactured by ABB is a fully digital system tailor-made for diesel engines and turbines. A number of hardware and software versions are available covering almost any application. These governors control either fully electric or electro-hydraulic actuators. The functioning of the governor can be monitored on a computer using software designed to allow intuitive operation utilizing a fully graphic interface. These governors are approved by the major marine classification societies.

Advantages

Compared to traditional Woodward type governors, electric governors offer many advantages. In addition to faster response, since they do not need a mechanical drive, they can be mounted remote from the engine; they are ideally suited for automation; offer smooth control of speed and fuel regulators responding dynamically to load conditions; controls loading and unloading, acceleration and deceleration and synchronised fuel pump operation.

Over Speed Trip

This device is provided to safeguard the engine against over speeding (due to slack governor response or failure) and works by activating the emergency stop solenoid to shut down the engine. The over speed sensing device initiates the shutdown signal to the engine, which is then acted upon by some other device on the engine to shut it down. The sensing device provides oil flow to a shutdown servomotor in the engine fuel system or to stop the flow of oil to a shutdown system when speed exceeds a given point. This shuts off the fuel to the engine either by deactivating the fuel pump plungers or by opening the pump suction valve. After tripping, the sensing device can be reset either manually or automatically once engine speed falls below the tripping speed to a predetermined level below the tripping speed. The manual reset type on the other hand, is equipped with a "lock-out" latch which engages after the over speed sensing device trips, preventing the device from resetting itself. An indicator shows when the manual reset knob is in a tripped position.

Both the reset speed and the tripping speed can be adjusted. Also, an external terminal shaft permits the unit to be tripped manually by a remote connection to the shaft. This could be electric, pneumatic or hydraulic. The system could be linked to operate from other input signals as well – like low lube-oil pressure or other vital engine system parameters. There is a provision to override the device during emergencies.

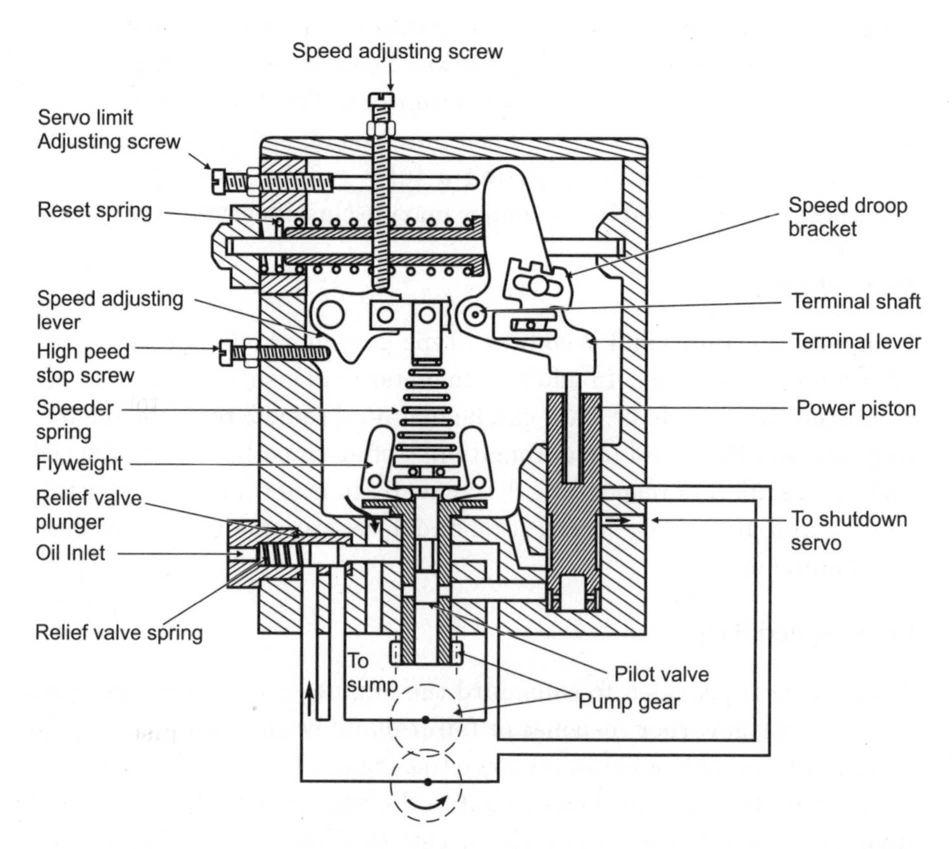

Fig. 3.17 Over speed sensing device.

A hydraulic over speed sensing and tripping device is shown in Fig. 3.17. The device uses an external oil source. Oil enters the relief valve inlet, flows down through the inlet port and reaches the pressure side of the pump. This internal oil pump builds up pressure until the relief-valve plunger is pushed to the left against the relief-valve spring. This uncovers the bypass hole in the relief-valve sleeve thus recirculating the oil through the pump. The normal operating pressure of the device is between 5 and 12 bar in addition to the inlet supply pressure. In the event of tripping due to over speed, the device will require an excess quantity of oil to supply to the shutdown servomotor. This will lower the oil pressure and the spring will move the relief valve towards the right. The recirculating passage gets blocked thus maintaining operating pressure, and the excess oil needed enters the pump through the inlet port.

The movement of the power piston is controlled by the pilot-valve plunger by controlling the flow of oil either to or from the area under the piston. In turn, the power piston controls the flow of oil to or from the shutdown servomotor in the engine fuel oil system. At normal engine speed, the pilot-valve plunger is under two opposing forces. The speeder-spring force tends to push the plunger down while the centrifugal force developed by the rotating flyweights tries to lift the plunger. In this equilibrium position, the speeder spring force holds the pilot-valve plunger down, relieving the oil pressure under the power piston to the drain. The reset spring, exerting pressure against the terminal lever holds the power piston down. In this position, the shutdown servomotor line is opened to the sump through the cavity around the flyweights. If the speed increases beyond the tripping speed, the centrifugal force of the flyweights overcomes the downward acting speeder-spring force and pushes up the pilot-valve plunger. As the plunger lifts, pressure oil flows to the underside of the power piston. The piston is pushed up, rotating the terminal lever against the force of the reset spring. Since the speed-droop bracket is connected to the terminal lever, as the terminal lever is rotated by the upward moving power piston, the right end of the floating lever is raised by the pin in the speed droop bracket. This tends to lift the speeder-spring, thus decreasing its downward force, and the flyweights move to their extreme “out” position. This feedback provides instant movement from normal to over speed position of the power piston. The power piston thus moves to the top of its stroke to the extent allowed by the terminal lever. In this new position, the power piston shuts off the drain line to the flyweight cavity and simultaneously establishing a direct connection between pump pressure and the shutdown servomotor.

A pressure gauge in the outlet line to the shutdown servomotor serves to provide external indication of trip and reset.

When the engine speed decreases to below the reset speed, the flyweights move inward and the speeder spring pushes the pilot-valve plunger down. Pressure under the power piston is relieved as the area under the piston is once again connected to the sump. As the reset spring rotates the terminal lever the power piston is pushed down, pump pressure is cut off, and the shutdown servomotor line is once again connected to the sump through the passages in the governor case. Thus, the device is reset automatically. To allow manual tripping, an external terminal shaft is provided. Turning this shaft counterclockwise lifts the terminal lever. This allows a spring under the piston to force the power piston up. The shutdown servomotor is thus connected to pump pressure. The tripping speed is determined by the position of the speed-adjusting screw. It is screwed in to raise the tripping speed and screwed out to lower it.

❑

Chapter **4**

Engine Systems

Scavenging

Scavenging refers to the effective removal of exhaust gases from the cylinder on completion of combustion, and induction of fresh charge air for the following compression stroke. This is essential for number of reasons: to ensure that the fresh charge is not contaminated with residual exhaust gases and thereby contribute not only to good combustion but also keep the mean cycle temperature relatively low; to assist in cooling cylinder, piston and valves and thus reduce thermal stressing of engine parts. Efficient scavenging therefore helps maintain maximum performance and economy.

Scavenging in Two-stroke Engines

Since the cycle is completed in only one revolution, there is a very short time available for purging the cylinder of exhaust gases and recharging with a fresh air supply. The gas exchange process takes place while both the inlet and exhaust ports are open and the piston is close to the BDC. Charge air under pressure admitted from the inlet ports, uncovered by the downward moving piston, sweeps through the cylinder and expels the exhaust gases in front of it. Though it is the exhaust port that is uncovered first by the piston (resulting in blow down of the gases at the end of the expansion stroke), a certain amount of mixing of charge air and gases is unavoidable. This is minimised by admitting a large charge air in excess of the cylinder volume, the excess air getting expelled along with the last of the gases. Exhaust gas driven turbochargers, under-piston compression or electrically driven auxiliary blowers ensure that pressure in the scavenge air manifold is maintained higher than that in the exhaust manifold, creating a clear pressure drop between the two, to obtain the necessary flow through the cylinder for good scavenging. Scavenging is obtained by any of three basic methods as shown in Fig. 4.1.

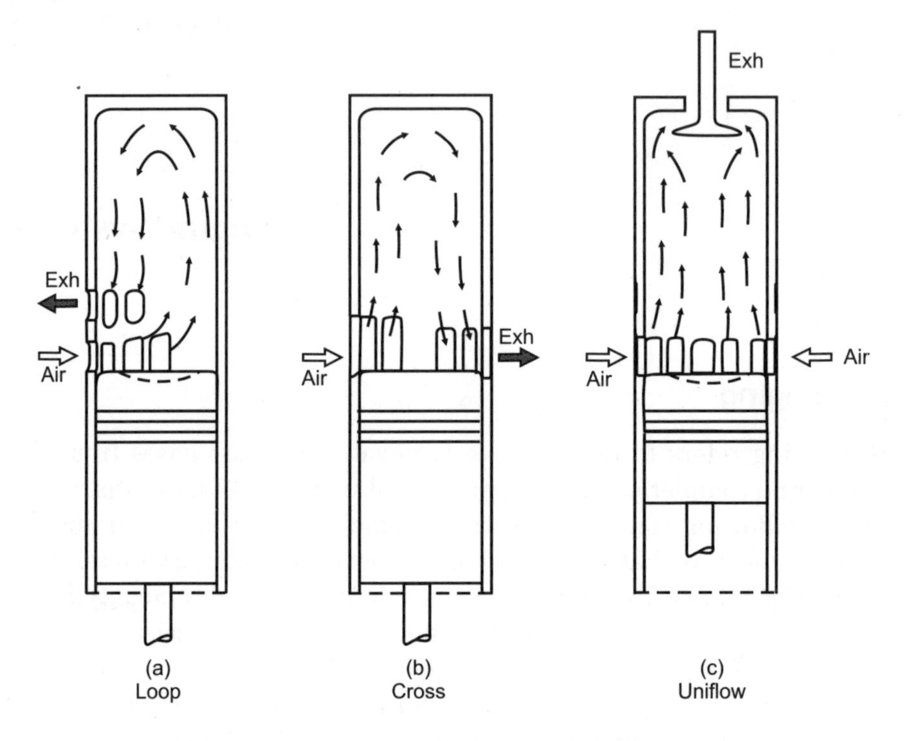

Fig. 4.1 Scavenging methods in two-stroke engines.

Loop Scavenging in which charge air flows across the piston crown and rises up in the cylinder on the opposite side, then loops back when it hits the cylinder head and finally moves down pushing the gases into the exhaust ports on the same side as the inlet ports. The piston crown is concave to assist in deflecting the gases upward as it flows across. The inlet port is angled in such a manner that the air is given a swirl during scavenging, thus creating the needed turbulence for good fuel-air mixing and efficient combustion.

Cross Scavenging in which charge air is deflected upwards as it hits the convex rim of the piston crown, raises up along the same side of the liner, then passes across the cylinder head to the opposite side and sweeps down to the exhaust ports on that side. The inlet ports are angled upwards to assist the upwards flow of the charge air. In both the above cases a long piston skirt prevents leakage of charge air into the exhaust manifold when the piston is at TDC. Scavenging efficiency is relatively low because of complex gas-air interchange.

Uniflow is a later development where air, entering the cylinder through ports around the full circumference of the liner at the bottom of the piston stroke, flows upwards right through the cylinder and exits through a large centrally positioned poppet-type exhaust valve on the cylinder head. This system offers the highest scavenging efficiency and the least mixing of air and gas. High temperature gradient between adjacent inlet and exhaust ports and on either side of the piston is also avoided. Due to the absence of exhaust ports cylinder liner is relatively simple in construction and its lubrication more effective. This system is mostly adapted for large, low-speed engines burning low-grade heavy fuels.

Latest engines have exhaust valves operated by a **hydraulic push rod** actuated by a servo-oil system receiving input signals of crank angle from analogue position sensors (*see* Fig. 3.4). There is one exhaust valve actuator for each cylinder. The electronically-controlled actuating unit gives complete flexibility in the timing and operation of the valves. Hydraulic servo-oil pumps maintain a pressure of up to 200 bar in the servo rail, from where oil is directed to individual actuating units through solenoid valves regulated by microprocessor control units, separate for each cylinder. The precise control of exhaust valve timing in response to varying loads ensures optimally excess charge air even at low loads thus reducing fuel consumption at low load levels.

Scavenging in Four-Stroke Engines

Scavenging is carried out with relative ease in four-stroke engines by the pumping action of the piston during inlet and exhaust strokes – by

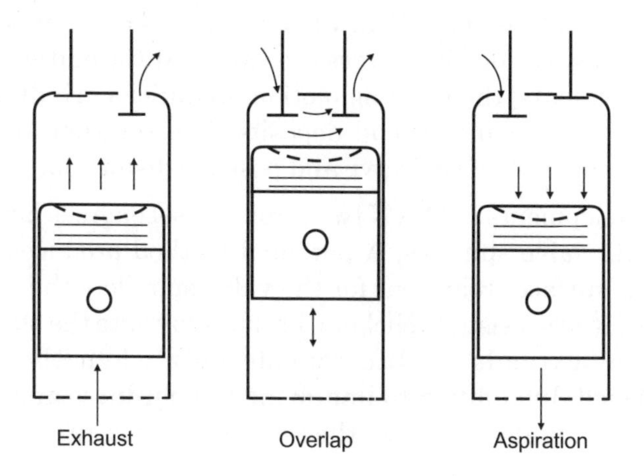

Fig. 4.2 Scavenging in four-stroke engines.

filling and emptying the cylinders – and by careful timing of the valves. This involves opening the exhaust valve before BDC at the end of expansion stroke and closing the inlet valve after BDC at the beginning of compression stroke and, providing for a higher period of overlap – up to 140°, as opposed to only 40° to 60° for normally aspirated engines – during which both the valves are open. It is also ensured that during the period of overlap the exhaust pressure is lower than the scavenge pressure. This results in a positive and more efficient scavenging with the least mixing of air and gas. The cylinder head mounted inlet and exhaust valves and their operating gear make the engine construction more complex though than two-stroke engines.

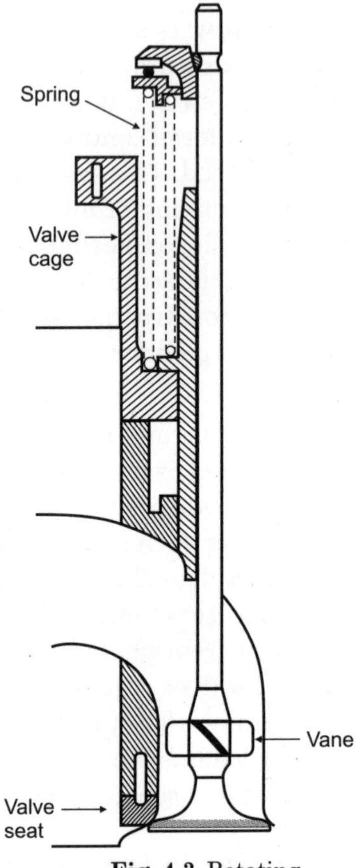

Fig. 4.3 Rotating exhaust valve of a four-stroke engine.

Fig. 4.3 shows a caged exhaust valve fitted with a rotating device that is employed in large medium-speed four-stroke engines. Vanes fitted on the valve spindle cause the valve to rotate as exhaust gases flow across. To facilitate the rotation the spindle is held in a bearing at its top end between the collects and the top spring carrier. The valve is therefore free to float when it is open and rotates in reaction to the exhaust gases flowing across the vanes. At the end of the exhaust as the valve reseats, the rotating motion produces a grinding action on the seat, thus removing carbon deposits. The rotation also prevents localised overheating of the valve and possible distortion.

New developments in M A N two-stroke diesel engines have stainless steel exhaust valve spindles. A patented method produces a resilient and durable surface treatment for the valve spindle – the welding of a high temperature-resistant Nickel-Chrome alloy onto the stainless steel spindle that extends its life considerably when the surface is work-hardened. In addition to improving the hardness and ductility of the spindle seat it also improves the seat's resistance to cracks.

Turbo Charging System

Since the amount of power an engine can develop is limited by how much fuel it can burn, and the amount of fuel it can burn is limited by the amount of air in the cylinders, it follows that to get more power more air is needed for combustion. This is achieved by compressing the air before it is charged into the cylinders. This compression increases the mass of the charge air and consequently, its pressure and density. This is what a turbo-charger does. A turbo-charger consists of a turbine run by the exhaust gasses on one end of a shaft and a rotary air compressor or blower at the other end. About 65 % of the energy available in the exhaust gases is recovered in this way. Turbo charging (also called supercharging) increases fuel efficiency considerably by squeezing as much energy as possible out of a given amount of fuel by burning it as completely as possible (while normal aspiration produces an air pressure of 1 bar turbo charging produces as much as 3.8 bar). The improved combustion, enhanced by the improved turbulence and scavenging due to supercharging, translates into a reduction in specific fuel consumption at all engine loads. To put it another way, given that for optimum combustion a stoichiometric air-fuel ratio of 1:14 is required, since a larger mass of air is available due to supercharging, more fuel can be burned for a given size of the engine allowing more power to be developed. Thus, this also translates into higher thermal efficiency due to the increased power output for a given size of the engine. This also results in a more powerful engine which is smaller in size requiring less space and with the added advantage of a reduced initial cost.

The efficiency of the system is further increased by cooling the pressurised charge-air thus increasing its density and consequently its mass. Another advantage is that the passage of cooled charge air has a cooling effect on the piston and cylinder liner leading to increased life and reliability of these parts.

There are two distinct methods of turbo-charging, which employ either the pulse system or the constant pressure system.

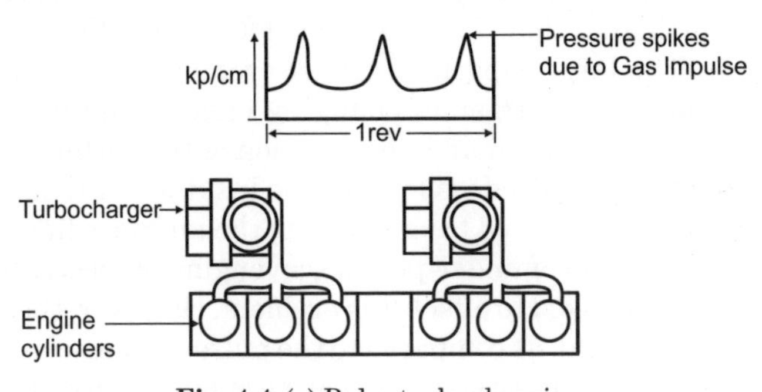

Fig. 4.4 (a) Pulse turbo charging.

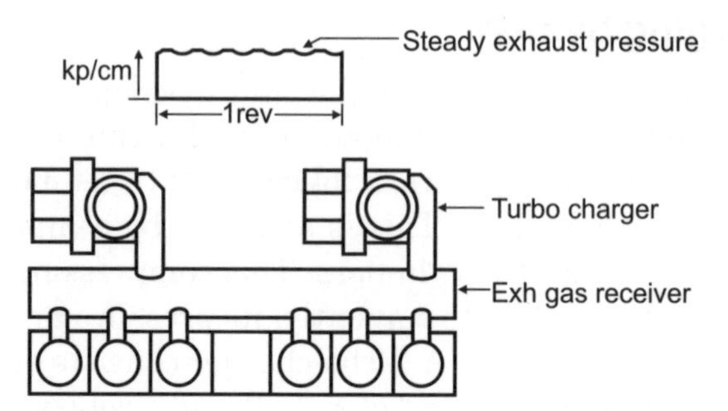

Fig. 4.4 (b) Constant pressure turbo charging – MAN engines.

Pulse System utilises the pulse or impulse energy of the pressure wave formed by the expulsion of exhaust gases from each cylinder. These waves travel through the exhaust manifold that connects a maximum of three cylinders to the turbocharger, Fig. 4.4 (a). The impulse energy gets converted to kinetic energy as the gases are ejected through the turbine nozzles. The high velocity jet of gases issuing from the nozzles drive the turbine blades. Pressure variation during a revolution is shown in the accompanying graph where the intermittent pressure spikes during each gas impulse from a turbocharger can be seen. The system provides quick build up of turbine speed during starting and manoeuvring since it is highly responsive to all engine running conditions. This makes the turbine speed vary by a few per cent during each cycle of the engine, whereas in the constant pressure system it is nearly constant. An auxiliary blower is sometimes used in the pulse system to ensure better acceleration while starting. Almost all marine four-stroke engines employ this system of turbo-charging.

Constant Pressure System has exhaust from all the cylinders discharging into a receiver where the pulses are converted into a steady pressure, Fig. 4.4 (b). This ensures that gas supply to the turbine is at a near constant pressure rather than intermittent pulses, as seen in the accompanying graph. The pressure energy increases with increasing engine speed and loading. Though at higher loads the turbine efficiency is high, at part-load conditions the performance is rather poor. Thus, at a higher bmep (higher than 7) this system delivers the highest quantum of charge air. The large capacity exhaust system however means slow response of the turbocharger to changing engine running conditions leading to slow acceleration and deceleration of the turbine. Therefore, engine exhaust at low speeds is found to be insufficient to maintain the required turbocharger speed thus reducing the air mass flow. Scavenge assistance, in the form of under-piston scavenging or electrically driven air blowers are employed to offset this problem. However, the number of turbochargers required, as compared to the pulse system, can be reduced in this system. Most large two-stroke marine engines run on the constant pressure system.

Turbochargers

About 25% of the energy released by burning the fuel is available in the engine's exhaust gases. Turbochargers recover as much as 65 % of this unutilized energy. Turbo-charging of four-stroke diesel engines began in the 1930s while it took another twenty years for two-stroke engines to be turbocharged. Two-stroke engines call for greater turbocharger efficiency than four-stroke engines to deliver the required larger air mass for scavenging while having comparatively less exhaust energy to drive the turbine. Therefore a notable difference between the two is the large size of turbochargers used with large two-stroke engines to handle the large air mass flows involved.

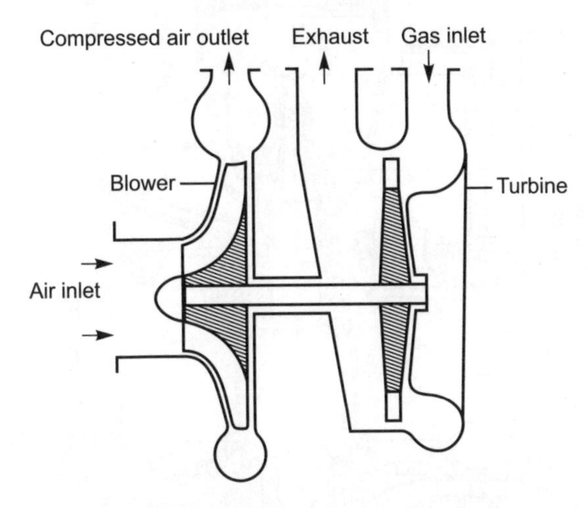

Fig. 4.5 Turbocharger, schematic diagram.

Turbochargers are single-rotor machines with a radial-flow centrifugal compressor driven by a single-stage axial-flow impulse turbine (Fig. 4.5). The radial-vaned compressor impeller and casing are of aluminum while nozzle ring and blades of the turbine are made of heat resisting steel and the inlet and outlet casing of high-grade cast iron designed for a working temperature in excess of 650° C.

Turbine: The fixed row of nozzle guide vanes consist of a number of converging channels that convert the pressure energy of the axially flowing gases into kinetic energy, imparting a high-velocity tangential momentum to the gases that are directed onto the single row of rotating blades, driving the turbine wheel. Lacing wire passing through holes in the blades dampen vibration. The inlet casing is designed for multiple-entry of gases for the pulse system while a single-entry casing suits the constant pressure system. Cooling water circulates through spaces around the casing.

Compressor: An air filter is mounted at the mouth of the compressor inlet casing. An inducer deflects incoming air into the impellers rotating at 10 -15,000 rpm. On the impeller discharge side a stationary diffuser plate directs air, forced out through the tip of the impeller by centrifugal force, into the volute casing converting the kinetic energy of the high velocity air mass into pressure energy. The casing acts as a collector conveying the air at constant velocity and pressure to the exit pipe. The casing is uncooled and the inlet passages are lined with a sound absorbent material to reduce noise.

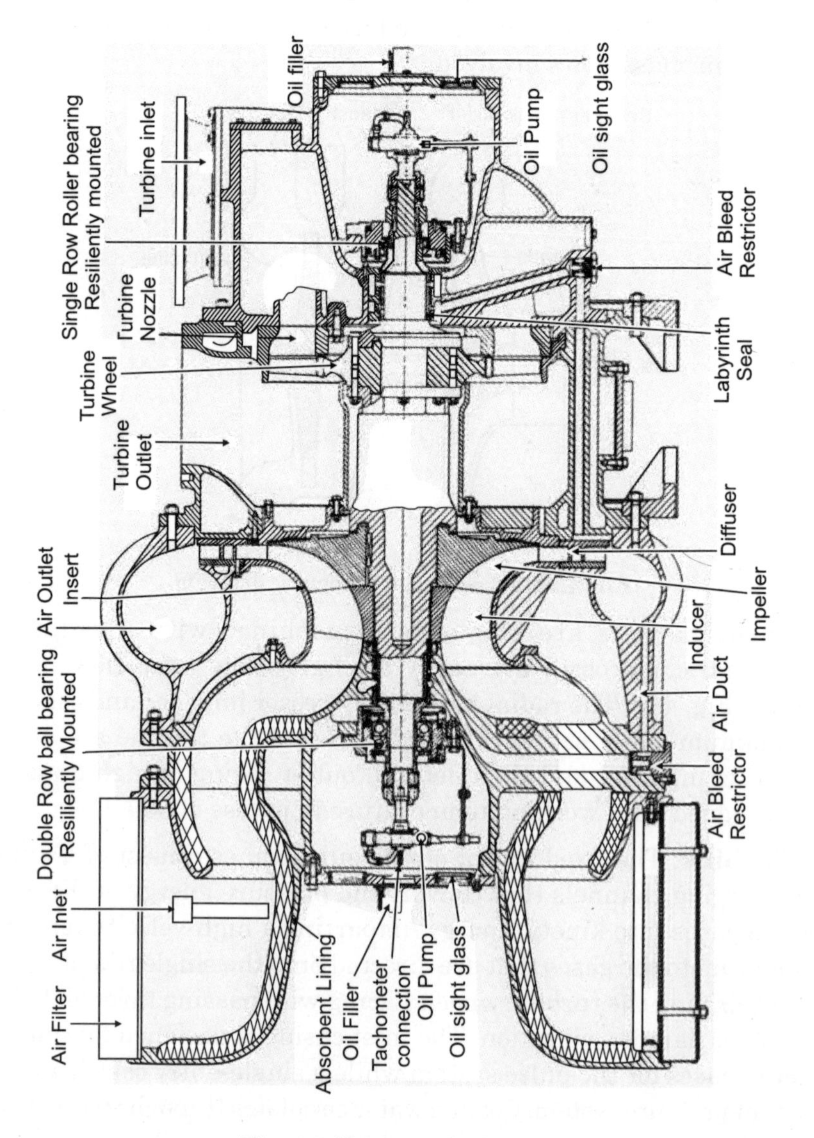

Fig. 4.6 Napier turbocharger.

As seen in Fig. 4.6, a double-row ball bearing supports the compressor end of the rotor shaft while a roller bearing is fitted at the turbine end, both resilient mounted, incorporating spring damping to reduce vibration. Both bearings are lubricated by a separate gear pump driven by the rotor shaft with its own oil sump contained within the ends of the casings (alternate designs have bearings supplied with lubricating oil from the engine system). Two labyrinth type seals, one fitted between the ball bearing and compressor and the other between turbine and its bearing, are sealed with air pressure from the compressor discharge through internal passages preventing oil leak into the compressor and exhaust gas leak into the oil sump.

A diagram showing the air and gas flow pattern of a uni-flow turbocharged two-stroke engine with hydraulically operated exhaust valve is shown in Fig. 4.7.

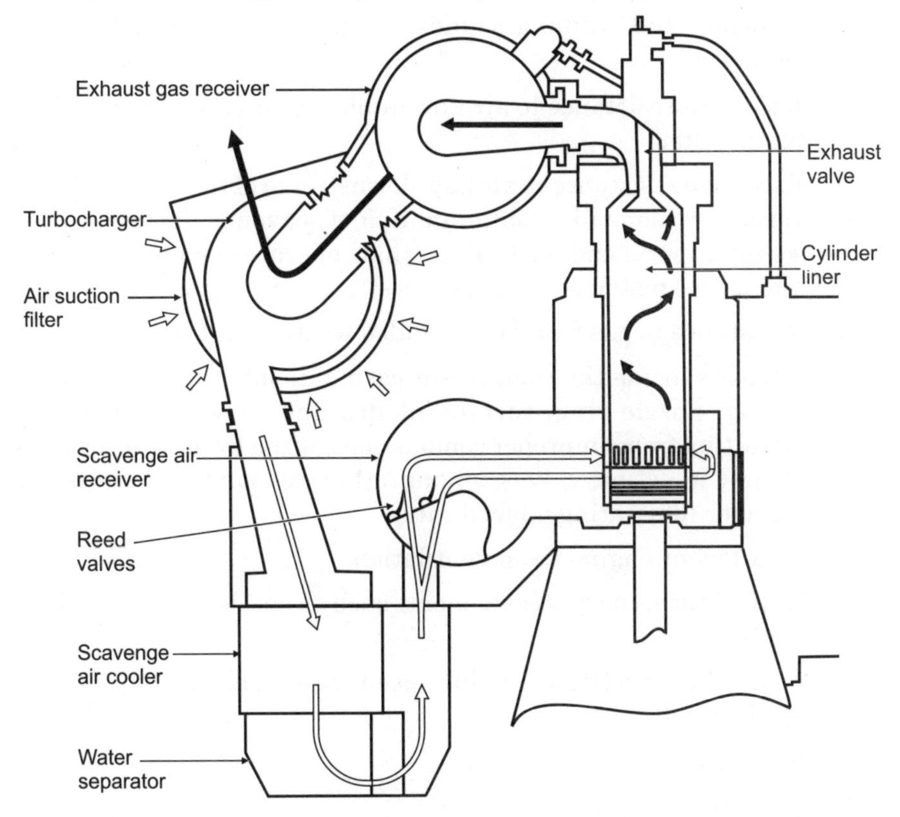

Fig. 4.7 Air/gas flow diagram.

Surge

Turbocharger surging is an occurrence resulting from periodic break down of air delivery at the compressor discharge followed by an immediate backward flow of air through the blower that will continue till the full air discharge is resumed by the blower. The break down develops when the air mass flow rate reduces considerably and correspondingly the pressure drop across the blower rises. That is, the pressure generated in the blower falls below the delivery pressure, leading to a break down of air delivery. Also if the pressure down stream of the compressor diffuser rises due to flow restrictions and goes higher than the pressure generated at the diffuser, this too will result in an immediate backward flow of air through the blower. Again, in a constant pressure turbocharged engine, with small numbers of cylinders or if the scavenge air manifold volume is too small, air pressure in the manifold will reduce toward the end of each induction process, and since the compressor maintains almost constant speed, the compressor may surge. At high engine loads, surging can lead to turbocharger rpm momentarily spiking by as much as 15%, which has the potential to damage its rotating elements. Characterized by loud howling and whining sounds, surging can be brought on by many events such as:

1. Excessive resistance to air flow at blower inlet due to fouled air intake filter.
2. Excessive resistance to air flow downstream of the blower, due to various reasons like contamination of scavenge ducting, fouled scavenge ports and reed valves (that separate scavenge manifold and under piston scavenge spaces).
3. Resistance to gas flow due to fouled waste heat boiler.
4. Deposits inside the compressor casing, inside turbine gas inlet casing, nozzle ring, turbine blades and scavenge air cooler (resulting from improper combustion caused by damage or wear of fuel injection valves among other reasons and excessive cylinder lubricating oil feed rate).
5. Insufficient engine room ventilation.
6. Wear of turbocharger components such as nozzle ring and turbine blades.
7. A sudden reduction of engine speed with corresponding drop in air demand.
8. Scavenge fire.
9. Hunting governor.
10. Rapid changes in engine load.
11. Air or water in fuel oil.

It is clear therefore that the best prevention against surging is careful attention to engine maintenance of which cleaning of the turbocharger is an important aspect. A temporary remedy is possible by reducing the engine speed and power. The relief valve on the scavenge air manifold could be opened or the auxiliary scavenge air blower, if provided, could be started to help the charger pass the critical point. If the problem persists, the turbocharger should be water washed and the air cooler cleaned. Also the gas passage from turbine to the funnel should be checked to make sure it is free. If surging is still persistent, then at the earliest possible opportunity the turbocharger turbine, cover ring, nozzle ring, compressor and diffuser should be inspected as described in the T/C manual and any fouling must be removed and worn components replaced.

Maintenance Procedures

As stated, internal working surfaces of the system can get fouled by adhering contaminants. Oil and dust drawn in from the engine room air foul the compressor while used cylinder oil, combustion products like carbon and ash foul up the turbine. Likewise, sulphur in the exhaust gases and salt in the air intake can deposit on the inducer blades and cause corrosion. The resulting hard deposits impede gas and air flow, cause local overheating and interfere with the dynamic balance of the rotating parts. This degrades the turbocharger performance apart from causing surging and undue vibration.

Water washing is employed to clear the contaminants. On the compressor side a small jet of clean water is injected into the air stream at the air inlet for up to 10 seconds that scours the internal surfaces clean. A jet of water sprayed into the gas flow at turbine inlet will soften the deposits sufficiently to be blown clear by the gas stream. Water is sprayed through an orifice for up to 20 minutes with the drains open to discharge excess water. To dislodge hard deposits chemical cleaning instead of plain water is sometimes preferred. For heavily contaminated machines, in cases where the turbocharger has not been cleaned for a long time, activated charcoal particles or natural kernel granules blown by compressed air is used instead. As a precaution against thermal shock, the engine speed is generally reduced to bring down the exhaust temperatures to between 300°C and 430°C before water washing on the turbine side. It is also important to ensure that the machine is fully cleaned as partial removal of deposits can result in rotor imbalance. Cleaning is most effective when carried out at high engine load. Though cleaning frequency is influenced by various factors like ambient air condition, quality of fuel, engine load and engine condition, the following is a general guideline: the turbine is water washed every 48 to 150 hours of running time while the compressor every 100 hours. Dry cleaning using compressed air and granules can be carried out once every 24 to 48 hours. Stubborn deposits which cannot be removed by the cleaning operation has to be removed manually during overhauls.

Charge Air Cooler

Compression will raise charge air temperature considerably, raising its specific volume, necessitating a cooler between the turbocharger and scavenge air manifold as shown in Fig. 4.8 (a). In order to remove fine entrapped water droplets from the air, a water separator is generally fitted after the air cooler as shown in the figure.

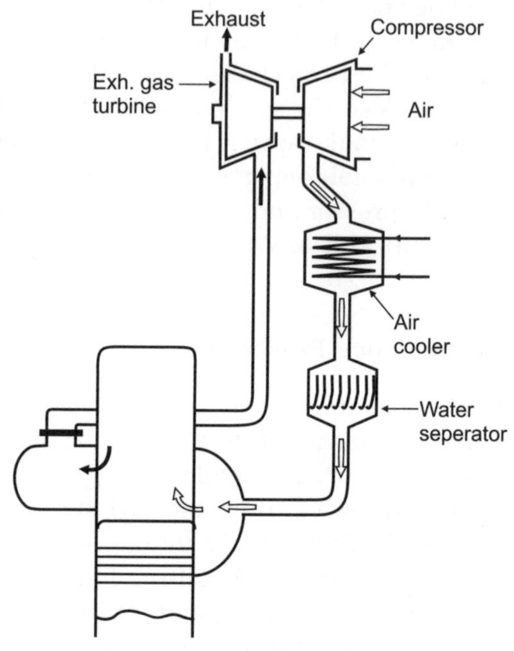

Fig. 4.8 (a) Arrangement of scavenge air cooler, water separator and air/gas flow.

Lowering the air temperature would increase air density, improving scavenging efficiency and the power developed. Additionally, scavenge air cooling ensures that the correct mass of air is supplied to the cylinder for efficient combustion and also that the exhaust gas temperatures are maintained within acceptable limits.

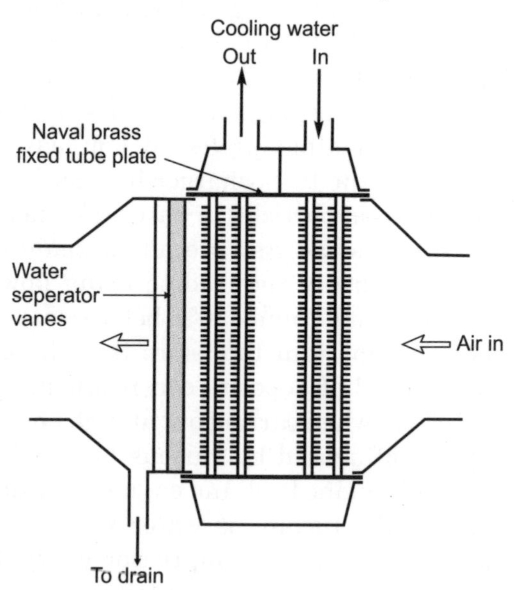

Fig. 4.8 (b) Radiator-type charge air cooler.

The water-circulated radiator-type air cooler, shown in Fig. 4.8 (b), is built up of a stack of copper-finned, aluminum-brass tubes secured to naval brass tube plates that carry cooling water inlet and outlet connections. While the inlet tube plate is fixed, the outlet plate is free to move in order to allow for expansion of the tubes. Air makes a single pass across the stack (to avoid high pressure drop) while cooling water makes two passes through the tubes (to maximize the cooling). Sacrificial anodes fitted inside the water boxes reduce corrosion from sea water*. In order to reduce the speed of the air stream and thus enhance cooling efficiency, air inlet is made divergent and the outlet convergent. Condensation of moisture in the air is bound to occur during cooling, especially if the air is under cooled to below its dew point. The condensate is removed through a drain fitted to the air outlet. However, in order to remove entrained water droplets from the air, a stack of water separator vanes is usually fitted after the cooler pipes as shown in the figure.

At dew point the entrained moisture in the air will condense into water droplets which can deplete the lube oil film on the cylinder liner walls. Grid type separators with angled baffles can remove as much 85 % water droplets from the air. To prevent excessive condensation air outlet temperature is maintained above 20°C. Cooling water flow is carefully controlled in order to maintain optimum charge air temperature. It is also important from the point of view of avoiding thermal shock to the cylinder liner.

Maintenance

An increase in air pressure drop across the cooler would indicate fouling of the fins while a high water pressure drop would indicate fouling of the water side (scarcely if fresh water is used for cleaning). Cleaning both the air and water sides would restore normal pressure drops. **Air side** is cleaned by injecting an air cleaning chemical into the air inlet side while the engine is running or blanking the air inlet and outlet when the engine is stopped and filling the space with the air cleaning solution and later rinsing it with clean fresh water. Protective clothing

**Corrosion is a spontaneous electrochemical process in which a metal reacts with its environment to form an oxide or other compounds leading to pitting and rusting. Anodes, which are alloys of Zinc, Aluminum and sometimes Magnesium, are attached to steel structures that need protection against corrosion. Anodes being a baser metal are more reactive and corrode in preference to the protected metal. Consequently, these anodes require renewal at routine intervals.*

must to be worn while handing the chemical. **Water side** is always cleaned with the engine stopped. The end covers are removed exposing the tubes which are then cleaned using a round long-handled wire brush and then washed with a water pressure jet. Sacrificial anodes should be inspected and replaced if necessary. Water side is generally cleaned only where seawater is used as the cooling medium. In the case of auxiliary engines, air coolers are removed from the engine and placed in a bath of cleaning solution, then wire brushed and rinsed with clean water. When immersed in the rinsing bath the cooler can be pressure tested to check for leaks.

Fuel Oil System

Marine diesel engines are designed to run on heavy residual fuel. This *heavy oil* is the residue after the lighter and more costly fuels and gases have been distilled out of the crude oil at the refinery. Consequently, these oils have a high viscosity and a high pour point, necessitating heating during storage and transfer – to make the oil pumpable. Steam heated coils in storage tanks and insulation and trace heating of fuel oil pipes ensure that oil temperature is maintained at least 5° C above its pour point to prevent solidification. However, for safety reasons, care is taken to ensure that the temperature does not reach its flash point – being always maintained about 20°C below this temperature.

Pour point is the lowest temperature at which oil will flow under prescribed test conditions.

Flash Point is the minimum temperature at which oil gives off flammable vapours that would cause momentary ignition when a flame is applied in a specified apparatus.

Viscosity of the oil is defined as the resistance to its flow, induced by internal molecular friction. Fuel oil viscosity decreases rapidly with increasing temperature. It is customary to indicate general viscosity at 40 °C. The temperatures for storing, transfer and injection into the cylinder are all based on the viscosity of the oil. For example, a higher viscosity tends to advance ignition by reducing fuel pump leakage. Hence, careful monitoring of the oil temperature is critical to ensure optimum performance of the fuel system, including correct atomisation by fuel injectors and life of the injectors and fuel pumps. A viscosity controller is invariably employed in the system to regulate heating so that fuel is maintained within its optimum viscosity range.

Calorific value is the heat released by the fuel upon combustion – the main constituents of the fuel that release heat upon combustion being hydrogen, carbon and sulphur. A typical marine heavy fuel oil could contain 85 % carbon, 12 % hydrogen and 3% sulphur by mass. The calorific value is measured by a bomb calorimeter where a small quantity of fuel is burned under controlled conditions and is expressed in k cal/kg. Gross calorific value for diesel oil is in the range of 10500 – 10700 and slightly lower for heavy oil at 10300 – 10450. Engine fuel consumption is found to be inversely proportional to calorific value, other things being equal. Modern marine diesel engines have typical fuel consumption between 165 – 170 g/kwh.

Density of the oil is required to calculate the quantity of bunker taken on board and is expressed in units of kg/m^3 (fresh water has a density of 1000 kg/m^3). Engine fuel consumption is measured in terms of weight though fuel is stored by volume. Volume of the oil is measured by sounding the bunker tanks and referring to the tank capacity chart against the depth, and that is multiplied by the relative density of the oil to get the weight. Relative density or **specific gravity** is the weight of a given volume of oil compared to the same volume of water, expressed as a ratio, both measured at a fixed temperature (usually 15°C). Diesel oil has a lower specific gravity compared to heavy oil and both are liable to vary slightly depending on the source of the crude oil and the refinery where it is processed.

Ignition quality is a significant property that determines the suitability of a fuel and essentially refers to the time taken for combustion to begin after the start of ignition. The shorter the combustion delay the higher the ignition quality, which translates into a smoother running engine and better cold starting characteristics. Ignition quality is expressed in either the **Cetane Number** or the Diesel Index (obtained by other parameters of the fuel). However, in large slow-speed engines fuel of low Cetane Number is preferred as otherwise a quick burning fuel can result in rather poor combustion and heavy carbonising of fuel injectors and high exhaust temperatures. The advantages of a higher Cetane Number therefore are obtained more in high-speed engines.

Carbon Residue indicates the relative propensity of the fuel oil to form carbon residues after burning. The Conradson test determines the residual carbon left after an oil sample is burned under specified test conditions. This value is an important consideration while burning heavy oil since it is directly linked to the extent of engine fouling.

A comparison of properties of diesel and heavy oil is given below.

Property	Marine diesel oil	Marine heavy oil
Specific gravity	0.84 to 0.88	0.92 to 0.99
Viscosity (Redwood No.1 in seconds)	37	860
Flash point (Pensky Martin, closed)	85°C Min.	90°C Min
Gross Calorific value (Kcal/kg)	10500	10330
Cetane No.	45 to 50	25 to 35

Heavy oil is transferred from the double bottom bunker storage tanks to the settling tank in the engine room by transfer pumps. Steam circulating through heating coils at the bottom of the settling tank heats up the oil to about 60°C allowing solid impurities like rust, sand and other abrasive contaminants to settle along with water and sludge (that forms when incompatible oils from different sources gets mixed or are blended together). This is periodically drained off through drain cocks. The oil is then purified by circulating it through centrifugal purifiers before being transferred to one of the two service tanks. Service tank temperature is also maintained in order to attain the appropriate viscosity before circulating the oil through the engine fuel system. For example, heavy oil with a viscosity of 1250 (redwood no.1 viscometer) at 40°C used in large two-stroke engines has a much reduced viscosity of 160 when heated to 80 ° C, rendering it fit to be circulated through filters and purifiers. Purifies form an integral part of the engine fuel oil system and generally two purifiers are used in series; one removes solubles, sludge and water while the other acts as a clarifier to remove solids. Prior to purification H.O. is heated to 95 ° C or more in a separate heater and D.O. to around 40°C. While one service tank is in use the other is filled from the purifier. H.O. service tank is maintained at around 90°C and D.O. at around 35°C.

Fig. 4.9 shows the fuel oil supply system for a large two-stroke crosshead engine. This is typical of any fuel system for a marine diesel engine operating on heavy oil. A supply pump draws oil from the service tank in use and delivers it at about 3.5 – 4 bar pressure to the engine circulating system. Next, a circulating pump boosts the pressure to about 10 – 12 bar and delivers it through a viscotherm-controlled heater and fine-meshed discharge filters to the engine. Viscotherm is an instrument

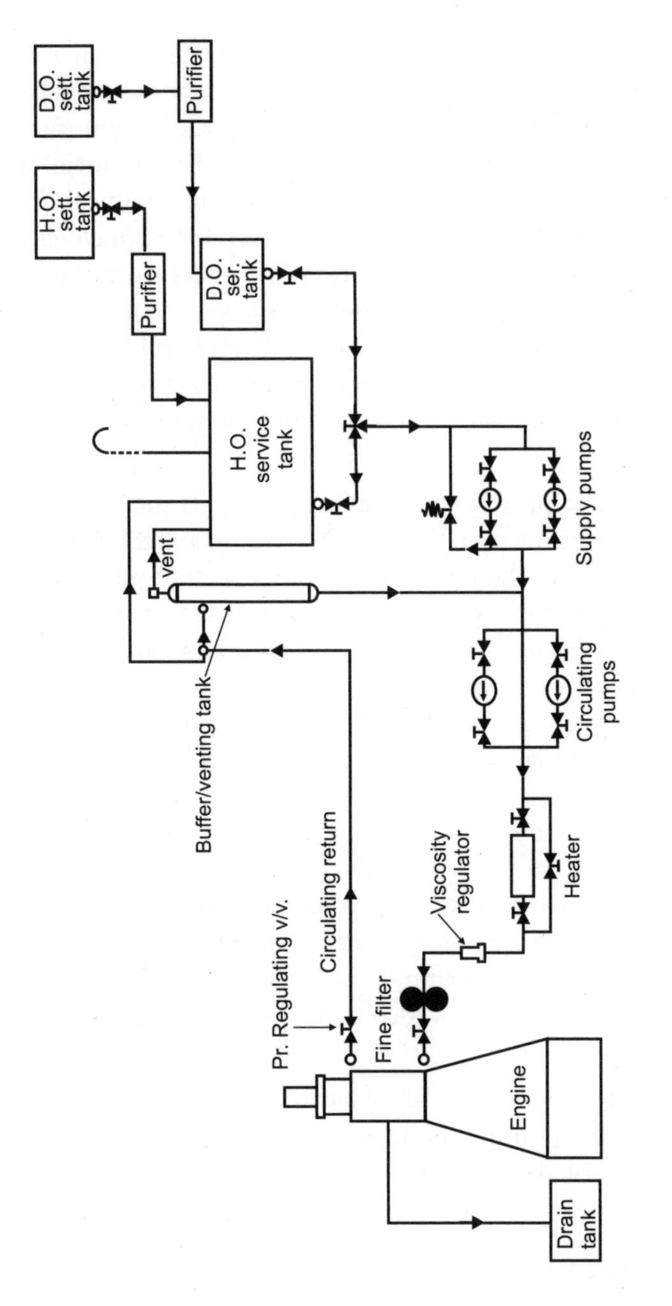

Fig. 4.9 Fuel oil system.

that measures the fuel viscosity at the heater discharge and operates the heater controls to maintain fuel viscosity within close limits. Re-circulated oil returning from the injectors and the fuel pump is directed to the return rail that discharges either back into the service tank or into a venting or buffer tank connected to the low-pressure side of the system from where the circulating pump draws the fuel oil.

A pressure-regulating valve connected in the return rail ensures a constant pressure at the fuel main by creating the required backpressure. This valve can be fully opened to by pass the engine fuel pump and circulate heated oil through the system to facilitate initial warming up. All tanks have arrangements for venting and draining. Pumps are in duplicate and have pressure release valves fitted to safeguard against excessive pressure build up. System low-pressure alarms and low oil level alarms fitted to tanks provide the necessary warnings. Other safety measures include quick-closing valves fitted to all tank outlets and the main engine inlet that can be operated from outside machinery spaces. Engine fuel pump is also fitted with remote actuators allowing emergency fuel shut off.

Diesel oil supply system similarly has a transfer pump to draw oil from the bunker tank and fill the D.O. settling tank. Suspended impurities and water settle to the bottom that is drained out periodically. Next, oil is circulated through a purifier that discharges into a service tank. From here it is directed to a three-way valve that has two inlet lines, one each from the diesel and heavy oil service tanks, and a single outlet. The valve permits only one type of oil to flow through to the outlet and thence to the engine. In conventional engines, diesel is used during starting and manoeuvring and later the engine is changed over to heavy oil, taking care to do it gradually, allowing sufficient time for temperatures in the system to stabilise. Similarly at the end of a voyage engine is changed over back to diesel oil to allow manoeuvring and to ensure that the system is flushed with D.O. before stopping (modern electronically controlled engines with common rail system however run only on heavy oil.) The D.O. service tank also supplies the auxiliary engines driving the alternators.

After long stoppage or after carrying out maintenance on the fuel system, the fuel pump, fuel lines and the injectors must be primed and the system purged of air to ensure that the engine fires immediately on starting.

Fuel Injectors

There are three important parameters that determine the quality of combustion of the injected fuel, namely, **atomisation** or breaking up of the fuel into minute droplets, **penetration** or the distance the atomized fuel travels into the combustion space and **turbulence** or the combined movement of the fuel and air inside the combustion space before combustion begins.

Atomized fuel has a high surface area exposed to the high air temperature that causes rapid evaporation and mixing. This is governed by the size of the injector nozzle holes and the difference between the fuel injection pressure and the compression pressure of air in the cylinder.

Good penetration is necessary to ensure that the fuel penetrates into the whole of the combustion spaces and mixes well with the compressed air. This is governed by the number of atomizing holes of the nozzle, their position and the injection pressure. The hole size and their downward angle is such that a symmetrical, conical spray pattern is produced.

Thirdly, good turbulence improves fuel air mixing for effective and rapid combustion. This is dictated by the fuel spray pattern, apart from the swirl imparted to the air by the shape of the piston crown and by the angle of the scavenge air ports.

Hence, for optimum combustion, fuel must be atomized and directed in the proper direction with the correct velocity. It is the function of the fuel injector to accomplish this.

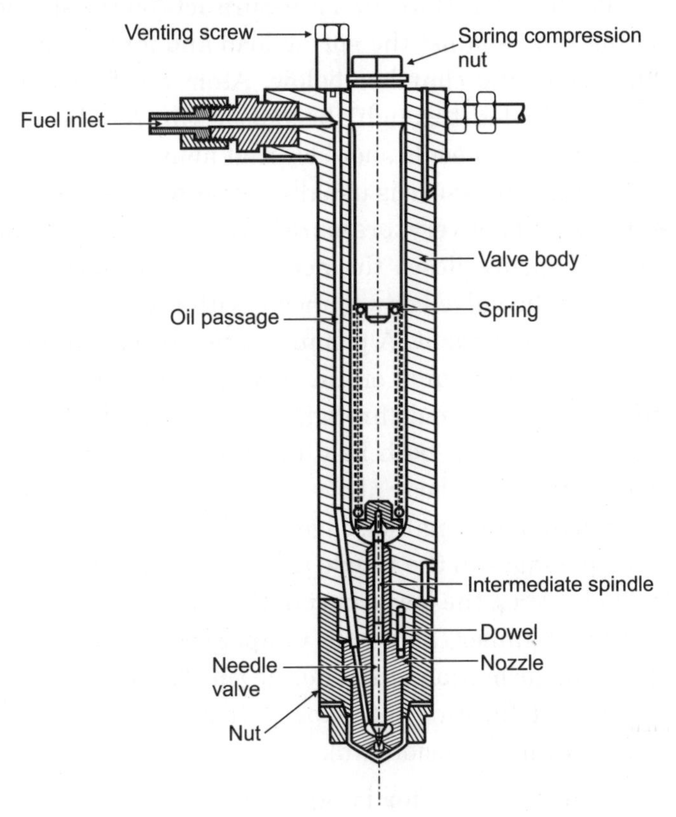

Fig. 4.10 Fuel injector.

Marine fuel injectors are mostly of the hydraulic type. Their design may vary according to the engine manufacturer. A typical hydraulic lift marine fuel injector is shown in Fig. 4.10. The lower face of the injector body and the upper face of the nozzle are precisely ground in. Similarly, the conical bottom tip of the needle valve and the beveled nozzle seat just above the atomizing chamber are also carefully ground in to form a mitre joint. These surfaces form an effective seal against the high injection pressure and ensure that there is no leakage of fuel. The needle is lapped into the bore of the nozzle, which is secured tightly on to the injector body by the retaining nut and aligned by a dowel pin that engages both. The needle is held down against its beveled seat by a spring pressing down through a valve spindle above. The spring is held in place between two guide washers, the lower one sitting on top of the valve spindle and the upper one touching a compression screw (spring tensioner) above, that helps adjust the spring load acting on the needle and hence the injection pressure. Matching drilled holes in the valve body and the nozzle direct high pressure fuel supplied to the injector by the fuel pump into the spindle chamber. Here, fuel pressure acts on the shoulder of the needle pushing it up against the spring load and forcing high pressure fuel into the atomizing chamber below. Atomized fuel is discharged through orifices drilled through the nozzle tip. Needle lift (usually 1 mm) is limited by the needle shoulder that abuts the lower face of the valve body. Injection pressure is usually set between 140 bar to 250 bar or more. At the end of delivery, pressure in the atomizing chamber drops sharply and the spring closes the needle valve abruptly. A threaded hole on the top of the valve body connects with the drilled fuel passage and the spring compartment. A priming screw threaded into this hole when loosened allows purging of air. Any fuel leaking into the spring compartment is also drained through this passage. In larger engines the nozzle is in two parts, both held together tightly by the retaining nut and against the valve body above. In some designs cooling water is circulated through additional passages drilled through the valve body and matching passages in the nozzle that reach almost to the tip of the nozzle. This helps keep the lower part of the nozzle, which is exposed to the combustion chamber, at a safe temperature below 200° C, thus preventing carbon formation at the nozzle tip. The water is maintained at 92° C to prevent flashing into steam. It is generally a closed loop system, with its own expansion tank.

Though ideally the injector is placed in the centre of the cylinder head (into a pocket and forming a gas tight seal at the lower landing and secured by studs and nuts), in large two-stroke engines with uni-flow scavenging system, where the exhaust valve is located centrally on the cylinder head, two or three injectors are placed symmetrically around the exhaust valve all supplied from a common distribution connection to inject equal quantities of fuel simultaneously.

Fuel Pumps

It is the function of the fuel pump to supply an accurately measured quantity of fuel oil to the injector of each cylinder at the precise time and at a controlled rate. Since the timing of injection is crucial, cams mounted on a camshaft and driven by the crankshaft are used to operate individual fuel pumps, one for each cylinder.

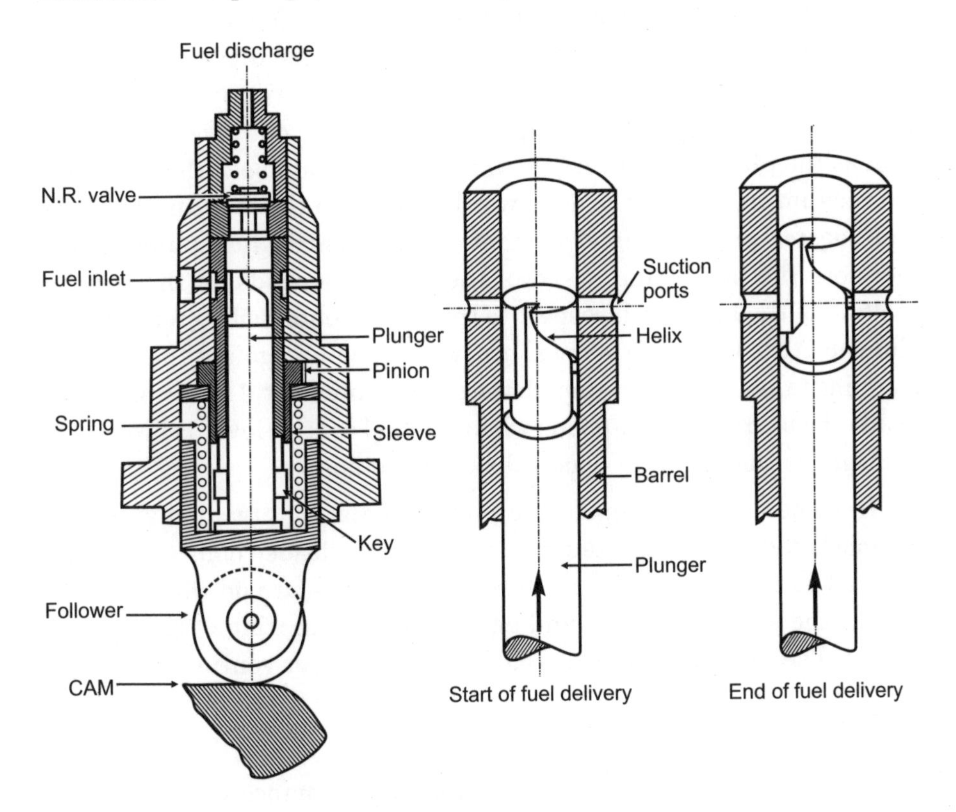

Fig. 4.11 Fuel pump.

Fig. 4.11 shows a jerk type fuel pump that is used widely in Sulzer marine diesel engines. The cam operated plunger works inside the barrel into which it is precisely lapped in. The helical spring pressing down on the cam follower ensures that the plunger, which is attached to the inside of the follower cup, returns down as the cam follower rides the base circle. The plunger has a helical groove machined into it along with a vertical groove and another annular groove below it. The barrel has two ports drilled into it – one suction and the other a spill port connected to the fuel oil suction outside the barrel. Both the ports are fully uncovered when the plunger reaches its lowermost position as it reciprocates inside the barrel following the rotation of the camshaft. The plunger is keyed into a sleeve around the barrel, which has a pinion machined on its top rim. The pinion meshes with a rack that is connected to the engine governor. Lateral movement of the rack rotates the sleeve – and the plunger – with respect to the barrel.

As the plunger moves upward, injection will take place the moment the top edge of the plunger shuts off the spill and inlet ports and the fuel above is trapped and then pumped upward. Injection stops as soon as the helical groove on the upward moving plunger uncovers the spill ports – and the pressure above the plunger drops abruptly as oil escapes into the spill port through the groove. During the downward stroke as soon as the plunger uncovers the inlet port fuel oil floods the barrel, by the vacuum created by the downward stroke and the pressure from the fuel booster pump. It is clear that the amount of fuel pumped and therefore the power and speed of the engine depends on the relative position of the helix with respect to the barrel, obtained by rotating the plunger. Also, when the vertical groove is lined up against the ports no pumping takes place even though the plunger might still be reciprocating inside the barrel. This happens when the fuel control lever is brought to zero and as the injection ceases the engine stops.

In some designs to make the plunger balanced, two helix grooves are machined on its opposite sides. Also, a spring-loaded non-return valve is fitted at the pump delivery to reduce the pressure on the discharge side at the end of injection and ensure positive seating of the injector needle. The discharge side also carries a priming or venting screw.

Fuel Pump Performance

The most important aspect of fuel pump performance is the injection timing and it is crucial that timing is maintained closely. There are number of factors that influence timing and tend to cause a variation.

It can be seen that while the end of injection can be controlled by rotating the plunger relative to the barrel, beginning of injection always occurs at the same point when the top edge of the plunger blocks off the spill port. However, due to wear of the plunger and barrel on account of abrasive particles in the fuel, and also erosion of the edges of the helix, the injection timing can begin to vary. Secondly, if the engine were to run on partial loads for a prolonged time, combustion may become inefficient leading to fouling of piston, liner and exhaust ports. This happens because at low loads the injection may be completed even before commencement of ignition and combustion. This delayed ignition can cause intense raise in pressure when combustion does occur leading to **diesel knock** and consequent thermal and mechanical stresses (for example, below 45 % engine load beginning of injection needs to be linearly retarded to maintain optimum performance). Additionally, the maximum firing pressure of the engine will reduce linearly as the power of the engine is reduced. This reduction of firing pressure will result in a reduction of the thermal efficiency of the engine and in turn increasing the specific fuel consumption. Again, different qualities of fuel (with different ignition properties) may require advancing or retarding fuel injection timing. Even elongation of the camshaft drive chain will require compensatory adjustments to the cam position. All these conditions will necessitate correction of fuel pump timing in order to maintain high efficiency and optimum fuel consumption at all operating conditions. This is achieved by one of three methods: 1. rotating the cam on the camshaft thus altering the angular position of the cam peak relative to the crankshaft, 2. raising or lowering the pump plunger with respect to its follower thus varying the plunger position relative to the inlet/spill ports, obtained by turning a plunger height adjustment screw between the plunger bottom and the follower (lowering the plunger will retard injection and rising it will advance injection) and 3. raising or lowering the barrel relative to the plunger by moving the pump casing on its mounting as in Sulzer engines or with the help of a barrel height adjusting ring connecting the top of the barrel and the pump casing as in B & W engines (lowering the barrel will advance injection while raising it will retard the injection timing). All the three methods however suffer from one drawback: they can be carried out only when the engine is stopped.

Developments in fuel pump design overcame this problem by making both the beginning and end of injection infinitely variable and automatically while the engine is running. This **variable injection timing (VIT)** is achieved by two methods: 1. by moving the pump barrel vertically up or down relative to the plunger, 2. controlling the pump delivery by suction and spill valves.

1. Jerk Type VIT Fuel Pump

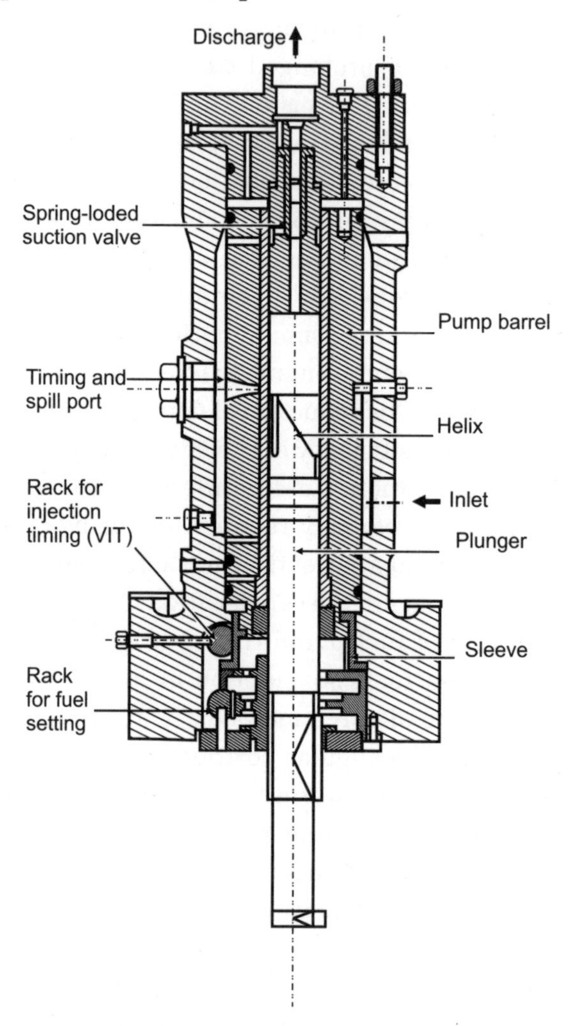

Fig. 4.12 M A N B & W Fuel pump with VIT.

The bottom of the barrel has a coarse thread cut into it which engages with a threaded sleeve called the timing guide and rotated by a toothed rack. The rack is connected to the engine governor. Movement of the rack and hence the vertical movement of the barrel will alter the position of the spill port relative to the plunger and therefore the timing of the start of injection. This allows collective adjustment of pumps as well as individual adjustments where required. Thus the pump has the normal fuel control rack that rotates the plunger to control the quantity of fuel injected (by varying the end of injection) and an additional timing control rack for varying the start of injection (Fig. 4.12). By incorporating pneumatic controls and a VIT servo system to actuate the rack even small adjustments required to allow for fuels of varying ignition qualities and changes in the camshaft timing due to chain elongation can be made.

2. Valve-controlled VIT Fuel Pump

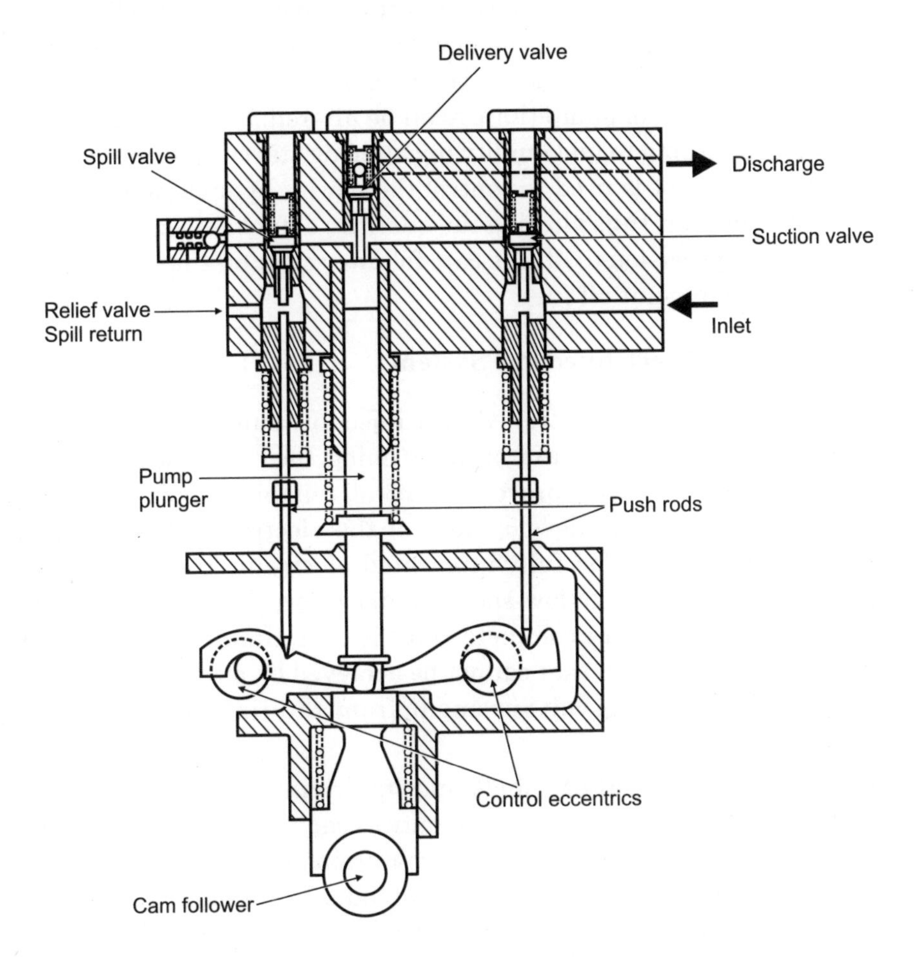

Fig. 4.13 Valve-controlled fuel pump of Sulzer engine.

These pumps are used in large Sulzer RTA engines and use suction and spill valves to control the quantity of fuel. Referring to Fig. 4.13, ends of two pivoted levers engage into an annular groove machined into the lower part of the plunger. One lever, pivoted in the middle, carries a push rod at its other end that operates the spring loaded suction valve. The other lever, pivoted at its free end, actuates a push rod that controls the spring loaded spill valve. While the suction valve controls the inflow of fuel into the pump chamber the spill valve controls its out flow. As the plunger rises, the suction valve push rod moves down to close the valve while the spill valve push rod rises along with the plunger to open the spill valve. Injection begins when the suction valve closes and pressure rises above the plunger and ends when the spill valve opens and releases the pressure. A non-return valve on the discharge side ensures that the pressure in the delivery pipe to the injector is not released. The pump chamber is recharged during the downward stroke when the push rod lifts the suction valve off its seat. The pivots of the levers are eccentric and by rotating these, pump timing can be altered. Suction valve pivot is adjusted according to fuel quality (by altering start of injection) while the spill valve pivot is used to control the power, by making the valve open earlier or later thus altering the quantity of fuel delivered. All fuel pumps have a provision to lift the cam follower clear off the cam using a pivoted handle to shut off fuel.

Common Rail Direct Injection System

Although double valve controlled fuel injection pumps with variable injection timing offer a degree of variable timing, the variation so obtained is still limited. Efforts to improve the control, reliability and economy of engines have led to the development of the electronically-controlled common-rail system. The fully-integrated electronic control system, especially suited for large low-speed engines burning heavy oil (up to 730 cSt viscosity), offers complete control over timing and rate and pressure of fuel injection that cannot be achieved by purely mechanical means. It is thus a big advancement from the fixed timing of the traditional camshaft.

Instead of the usual camshaft and its driving gear, fuel injection pumps and all the related mechanical control gear, this system is equipped with a common rail in which fuel oil is maintained at high pressure by a supply unit, ready for injection into individual cylinders. The supply unit consists of a bank of fuel pumps driven through gearing from the engine crankshaft all delivering pressurised fuel oil into a common collector or accumulator from which two independent, double-walled delivery pipes lead upwards to the common fuel rail. The collector is equipped with a safety relief valve set to 1250 bar. The volume of the common-rail system and the supply rate from the fuel supply pumps are matched to get a very stable pressure (nominally 1000 bar) in the rail with negligible pressure drop after each injection. The number of pumps (from 2 to 8), their size and arrangement depends on the engine type and the number of cylinders. The pumps are driven by a camshaft, short and small in diameter – quite unlike the traditional engine camshaft. The supply pumps too are different. There is no sudden jerk action as in traditional fuel injection pumps but rather the pump plungers have a steady reciprocating motion, maintaining a near constant fuel pressure in the common rail. The rail, which provides a storage volume for the fuel, has provisions for damping pressure waves and is trace heated from the ship's steam heating system. A single rail unit can supply up to seven cylinders; two units being used for engines with more cylinders.

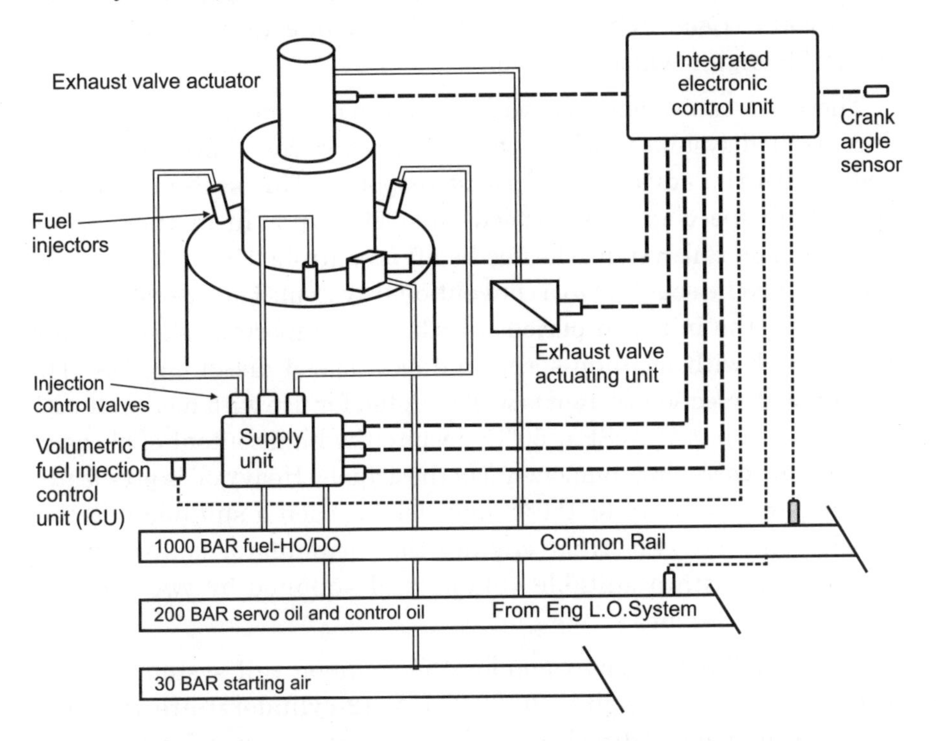

Fig. 4.14 Electronically-controlled common-rail system-RT-flex engine.

Fuel is delivered from the common rail to the fuel injectors through a separate injection control unit (ICU) for each cylinder. Thus, the functions of pumping and injection are separated. The ICU regulates precisely the timing of fuel injection, accurately monitors the volume of fuel injected, and sets the injection pattern – all accomplished by hydraulic servo-oil pumps, servo and control oil being drawn from the engine lubricating oil system. For each fuel injector the ICU has an injection control valve and an electro-hydraulic rail valve (a solenoid valve with an extremely fast actuation time), which receives control signals for the beginning and end of injection from the respective electronic unit of the control system. For increased life and durability, the rail valves are not energised for more than 4 milliseconds. This time is monitored and controlled closely by the engine electronic control unit. The precision rail valves are also isolated from the heated heavy fuel oil.

The modular electronic system, with separate microprocessor control units for each cylinder, receives input signal for the engine crank angle very accurately from two sensors driven from a stub shaft on the free end of the crankshaft. In engines with a cylinder head mounted exhaust valve, all the fuel injectors on one cylinder arranged around the valve normally act in unison, but can also be programmed to operate separately if necessary. Each cylinder has three fuel injectors except for the RT-flex 50 engine which has two.

There are two variations available in the common rail systems: the lift-controlled system in which the rail pressure constantly acts on the needle, with the advantage of a fast response. This system is mainly used in high-speed engines of speeds up to 5000 rpm. In another variant of the common rail system, the needle lift is controlled by pressure pulse under the needle seat like in a conventional system. The pressure exists only during the injection period, which is an important safety factor. This is used with medium-speed and slow-speed diesel engines. The common-rail injection system is well adapted for use with heavy oil with viscosities of up to 700 cSt at 50°C, despite the high content of abrasive particles and corroding components in these fuels. Heavy oil is pre-heated to a temperature of up to 150°C in order to reach a suitable injection viscosity. In large two-stroke engines the accumulator is divided into several units with a suitable volume and supplied by two separate high-pressure pumps.

Sulzer RT-flex engines employ the common rail system and the Sulzer RT-flex 96C engines (68,640 kW 12-cylinders) are the most powerful common-rail diesel engines in the world today (refer Fig. 2.9.3). As seen in Fig. 4.14, in addition to the common-rail fuel injection, the system provides full electronic control of other engine functions such as exhaust valve actuation and the starting air system.

Advantages

With the conventional injection arrangement of one fuel pump for each cylinder, a failure of one pump can lead to the loss of that cylinder and the consequent imbalance in engine torque necessitating drastic power reduction. In contrast, with the common rail system, in which all high-pressure supply pumps are grouped together and deliver to a common rail supplying in turn to all cylinders, the loss of a single pump has much less effect, the engine able to deliver full power even with one fuel pump out of action. Even in the event of further pumps hanging up, there would be only a proportional reduction in power.

Other benefits include a marked improvement in the following: low-speed operation, engine acceleration, balance between cylinders, load control, and longer times between overhauls – apart from better combustion at all operating speeds and loads, better fuel economy, lower exhaust, and a cleaner engine with fewer deposits of combustion residues on the internal parts. Additionally, there is low engine vibration, steady operation at very low running speeds, improved engine performance monitoring and near-smokeless operation at all speeds with precise speed regulation (Sulzer RT-flex engines are able to run steadily at ten per cent of nominal speed). An additional benefit is avoidance of fatigue failures of system components due to short duration pressure pulses (in excess of 100 bar) generated by conventional 'jerk' fuel pumps and consequent spillage of hot oil and fire hazards.

Maintenance

Fuel injectors need periodic and regular inspection, overhauling and pressure setting to maintain them in perfect condition. After dismantling, all parts should be washed in clean paraffin and dried. Use of cleaning rags and cotton waste should be strictly avoided since fluff adhering to cleaned parts can find their way into the nozzle holes and impede the spray, resulting in injection pressure rise and reduction in quantity of fuel injected. Valve needle and the nozzle-piece have to be handled with utmost care since these are precision fitted, matched parts (that always come in a pair). Scoring marks on the needle can be removed by lapping it inside the nozzle piece using fine buffing fluid then washing clean with paraffin. The parts have to be put back together with great care making sure that the dowel pin is not damaged, since a sheared dowel pin would mean 'no spray' or a misdirected spray. For this reason the spring tension must be completely relieved before tightening (or loosening) the nozzle retaining nut. If nozzle holes are enlarged (more than 10% of the original size) the needle and nozzle set should be replaced, else, it could result in bad atomization causing overheating or burning of the piston crown. After assembly the injector must be pressure tested on the test pump. Using the air vent screw to first prime the injector, the needle lifting pressure is then observed. The pressure adjusting screw that controls the spring load on the needle is used to set the correct lifting pressure stated in the engine manual. The screw is then locked in position, after which the lifting pressure has to be rechecked a few more times. Next, the nozzle tip is wiped clean and pressure re-applied, to about 10 kgf/sq. cm. less than the operating pressure. If the pressure holds for a few minutes the injector is leak proof. This is an important test since a leaky injector could result in air entering the fuel line between pump and injector and reducing pump efficiency leading to irregular running of the engine and if left unattended could cause the pump plunger to stick in its barrel or even seize. Further, if a fuel injector leaks fuel may enter the cylinder during a stroke and interfere with normal combustion. A low indicator card readily shows if a fuel injector is choked, leaking or sluggish.

Fuel pumps do not require rigorous maintenance as fuel injectors do. Comparison of indicator cards and exhaust temperatures of cylinders would indicate if fuel pumps are delivering correctly. If not, adjustments would be required. A compression card will show if the firing height (maximum pressure) is within permissible limits and a draw card will show if the mean effective pressure is alright. If the pressures are high the pump plunger should be lowered (by about half the amount the card is too high) in order to reduce the pump discharge. A low indicator card on the other hand will mean a sticking or leaking pump plunger or that the pump roller is not riding smoothly on the cam. Once the fuel pump is overhauled and defective parts replaced, the pump plunger height should be rechecked using the pump mandrel as directed in the engine manual and confirmed by checking the position of the fuel regulating rod with the starting handle in the stop position. The pump and fuel delivery line should be primed – with the roller on the base circle of the cam and using the cut-out lever.

Should a fuel pump need to be taken out of service while the engine is running, the fuel stop valve supplying the pump should be shut-off and the pump hung up (lifted clear of the cam using the cut-out lever) in order to prevent the plunger from running dry and seizing.

Problems in fuel oil system

Problems associated with fuel oil or equipments in the system will directly result in low engine power output and increased emissions due to poor combustion. Following is a list of potential problems:

- Poor quality of fuel (low calorific value).
- Low fuel viscosity.
- Worn fuel pump.
- High water content in fuel oil.
- Low fuel oil pressure.
- Blocked fuel filter.
- High pressure drop in the air cooler due to fouling.
- High cooling water temperature.

Fuel consumption of the engine is also related to the quality of fuel burned and the service condition of the fuel oil equipments. High fuel consumption could result from a combination of the following reasons:

- Poor quality fuel (low calorific value).
- High water content in fuel oil.
- Low fuel viscosity.
- Worn fuel pumps.
- Worn fuel injector nozzles.
- Late injection.
- Leakage in fuel filter.
- Fuel oil loss through oil purifier.
- Blocked air cooler.
- High scavenge air temperature.
- Fouled turbocharger.

Fuel Conditioning Module

To prevent some of the problems stated above, modern engines employ a fuel-conditioning module such as the Alfa Laval model that treats both heavy and diesel oils so that it meets the requirements of cleanliness, temperature, viscosity, pressure and flow rate as specified by engine manufacturers. Meeting these requirements will ensure not only superior engine performance but also decreased engine emissions. The system is compact and self-contained and is easily configured to meet the requirements of any make of engine or engine room layout. All the main components like pumps, filters, fuel heaters, pressure transmitters etc are mounted on a frame, complete with piping, valves, electrical cabling and other accessories. Since viscosity is the main control parameter, a viscosity transducer provides the necessary control signal and a flow transmitter provides information on fuel consumption. Pump starters and other process controllers are housed in a centrally located cabinet. Change over between heavy and diesel oil is carried out automatically through a ramp function to prevent temperature shock to injection equipments. When combined with Alfa Laval purifiers these modules afford complete fuel treatment from bunker tanks to main engine.

As shown in Fig. 4.15, the module consists of two stages, one, a low-pressure stage maintained at 4 bar and the other, the high pressure stage maintained between 6 and 16 bar depending on the type and make of the engine. The two stages are pressurised by a booster system that uses viscosity as the primary control parameter.

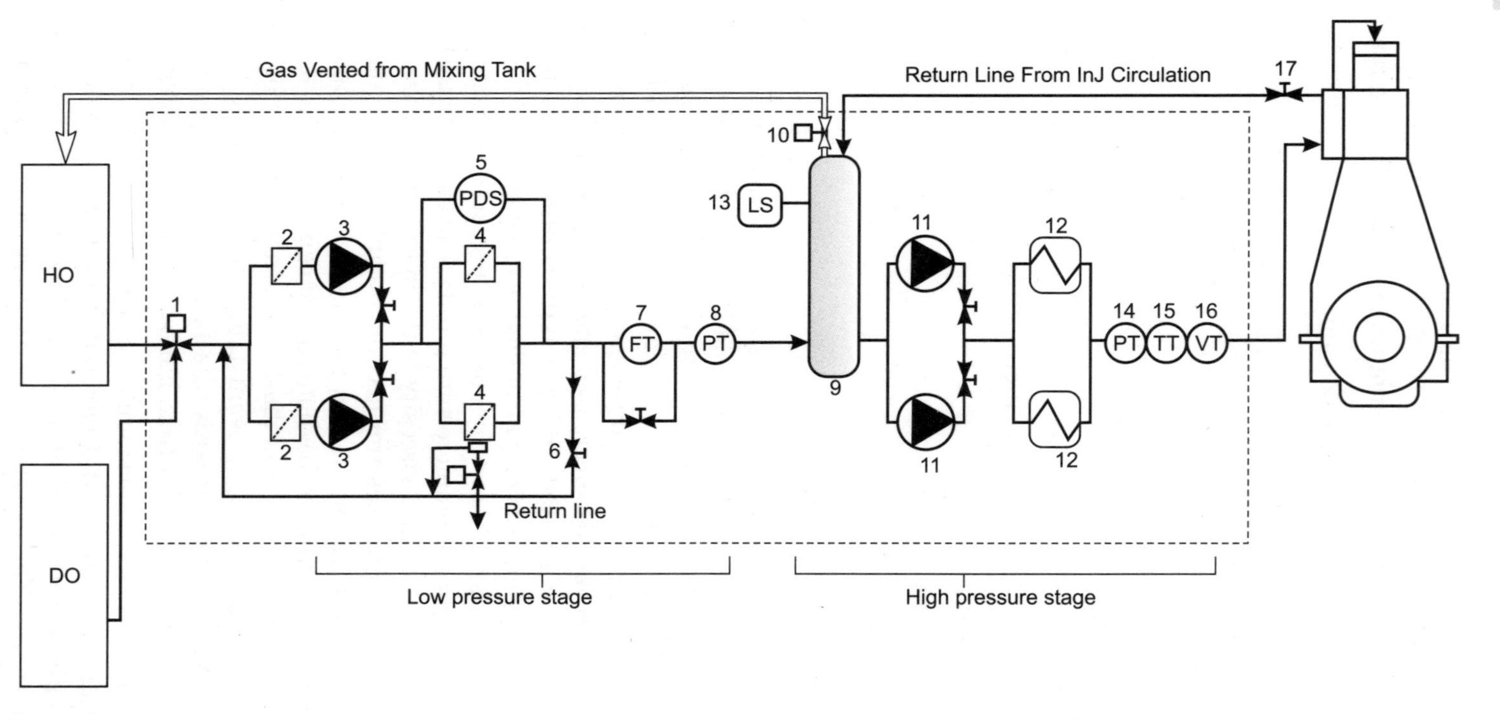

Fig. 4.15 Fuel Conditioning Module.

The low-pressure stage consists of a supply pump (in duplicate, so that if one fails the other starts up automatically), an Alfa Laval automatic filter backed up by a manual bypass filter and a flow transmitter to monitor fuel consumption. Fuel enters the module through the three-way change over valve 1, then passes through the strainers 2 and reaches the suction of the supply pump 3. The 4 bar pressure developed by the pump ensures that there is no gasification and cavitation due to the high fuel temperatures, between 120 –150°C. The pressurised oil then passes through the automatic filter 4. Continuous automatic back flushing and disc type filter elements ensure high efficiency and minimum maintenance. The filter pressure-drop switch 5 triggers automatic back flushing. A return line from the filter outlet connects to the pump suction via a pressure control valve 6 that not only helps regulate the supply pressure but also allows fuel to recycle within the low-pressure section so that the flow of fresh fuel entering via the three-way valve exactly matches the fuel consumption of the engine. The flow transmitter 7 and the supply pressure transmitter 8 generate flow-rate and pressure signals respectively and transmit them to the controller. Next, fuel reaches the mixing tank 9 where fresh fuel from the low-pressure stage is mixed with hot fuel returning from the injector circulation system of the engine. A level sensor 13 triggers an alarm in case of an abnormally low level due to any reason. Gases that accumulate in the mixing tank are automatically vented back to the fuel service tank via the automatic de-aeration valve 10. From the mixing tank fuel enters the high-pressure section. The flow rate in this section is always maintained higher than the actual fuel consumption rate in order to prevent fuel starvation at the injectors – both pressure and flow rate being set according to the recommendation of the engine manufacturer. The high-pressure stage mainly consists of the circulation pump 11, the fuel heaters 12 and pressure and temperature sensors and the viscosity transducer. Again, the pumps are in duplicate and should one fail the other starts up automatically triggering a 'pump shift' alarm. The plate-type heat exchanger 12, designed for pressures up to 16 bar and temperatures up to 170°C, uses special high temperature gaskets and steam as the heating medium. The pressure sensor 14 and the temperature sensor 15 transmit respective signals to the controller. The viscosity transducer 16 measures the viscosity of the fuel and compares it with the value set by the engine manufacturer and any deviation is corrected by the controller, which adjusts the flow of steam to the heater by actuating the electrical valve on the steam supply line. Pressure in the system is maintained by a pressure control valve 17 located in the return line after the injectors and the excess fuel is returned to the mixing tank.

The controller monitors and controls the functions of the system and displays process values such as viscosity, outlet temperature and instantaneous fuel consumption rate. The system automatically changes over from DO to HO using a temperature ramp function that actuates the three-way changeover valve. Fail-safe functions include, apart from pump changeover, shifting from viscosity control to temperature control or vice versa. There is also a provision to changeover to diesel should there be a heater failure.

Fuel Monitor

Water contamination can occur due to poorly maintained fuel supply system, condensation in fuel tanks, inefficient fuel purification or poorly maintained fittings leading to water ingress. As already mentioned, water contamination of fuel oil can cause a series of problems such as corrosion, accelerated component wear, microbiological growth and loss of power. These problems can be avoided by continuous online monitoring of fuel oil for presence of water. EESIFLO DACS SYSTEM is such a monitoring device that is generally installed with Mitsubishi and Alfa Laval oil purifiers for control of desludge operation and to trigger an alarm in the control room when water content exceeds a set limit. The fast response of this real-time fuel monitor means water contamination in any form at the purifier outlet is picked up immediately, except during start-up sequence and during purifier desludge cycle.

Engine Lubrication

The function of lubrication is:

1. To maintain an oil film between mating surfaces in relative motion so that metal-to-metal contact is prevented.
2. To reduce friction and therefore wear.
3. To prevent corrosion.
4. To carry away frictional heat.
5. To protect against impurities like dust, dirt and moisture.

A lube oil film is maintained in one of two ways: 1. by **hydrostatic lubrication**, where pressurised oil is injected in between two mating surfaces, like in a crosshead journal bearing (oil pressure tends to 'float' the crosshead pin). Here lubrication efficiency is governed by the oil supply pressure and the surface area between journal and bearing.

2. by **hydrodynamic lubrication**, where oil pressure is generated due to the relative motion between two mating surfaces, like a sleeve and the rotating journal, where oil is carried along the journal and forms a continuous unbroken film between the two thus preventing metal to metal contact. The factors that determine the effectiveness of this type of lubrication are the oil viscosity, speed of rotation of the journal in the bearing, the pressure between the two (on account of the load acting on the journal) and, the clearance between the two. The last mentioned has an affect on the oil flow through the system, since flow depends not on the supply pressure alone but also on the resistance in the whole circuit, dictated by running clearances*. If lubrication becomes insufficient and oil film thins considerably allowing partial metal-to-metal contact, the phenomenon is called **boundary lubrication**.

Lubricants have a low coefficient of friction; therefore to further improve lubrication they are blended with additives known as friction modifiers that chemically bind to metal surfaces to reduce surface friction. Other additives like corrosion inhibitors form chemical bonds with surfaces to prevent corrosion and rust. On account of their high specific heat capacity lube oils transfer heat effectively, the amount of heat carried away being dependent on the flow rate. Typically, lubricants also contain detergent and dispersant additives to assist in carrying debris and contaminant to the filter from where they are removed periodically when the filters are cleaned.

Lube Oil Additives

Marine lube oils are generally pure mineral oils derived from crude oil by hydro-cracking and solvent extraction processes. To impart performance characteristics, this base oil is blended with additives, such as:

- Viscosity index improvers – organic polymers added to maintain viscosity as constant as possible in spite of temperature variation.
- Anti-oxidant inhibitors – amines and phenols added to crankcase oils to resist oxidation
- Metal deactivators

** A thumb rule for bearing clearances allows one thousandth of an mm per mm of shaft diameter. However, for main engine main bearings three-fourths of this amount is taken as acceptable. Half the amount or lower would mean running the risk of 'wiping' of the white metal.*

- Corrosion inhibitors – alkaline additives like hydroxides added to neutralise sulphur compounds in fuel
- Rust inhibitors
- Anti-foaming agents – silicones added to prevent foaming due to entrained air in the oil. Foaming can lead to break-down of oil film in bearings.
- Friction modifiers
- Extreme Pressure reagents – organic compounds added to hydraulic and gear oils to maintain oil film under severe load conditions.
- Demulsifying / Emulsifying agents – polar compounds that emulsify without affecting lubricating properties.
- Detergents and dispersants – soaps and compounds added to cylinder oils to hold carbon particles and other deposits in suspension so they are easily conveyed out of the cylinder.
- Oiliness or wetting agents – fatty acids or chlorinated wax added to reduce friction and wear.
- Anti-bacterial agents – biocides added to prevent growth of bacteria.

Lubrication system and equipments are essential components of engines that provide and apply controlled amounts of specific lubricating oils at the proper temperature, viscosity, flow rate and pressure to various parts of the engine and other pieces of equipments to ensure their reliable and efficient operation. Thus, there is cylinder oil, crankcase oil, stern tube lubricants and trunk piston engine oil. In addition, a solid lubricant like grease is used where it is impractical to use oil. Grease is blended with molybdenum disulphide to withstand very high temperatures of up to 350°C.

Lube oil system is made up of oil storage tanks, oil sumps or drain tanks, pumps, filters and heat exchangers. Instrumentation to provide readings of the flow rate, temperature and level of the oil and alarms to warn of improper oil levels and pressures and temperatures form the rest of the system. While the main engine is provided with a centralized lubrication equipment with separate electrically driven lube oil pumps, the auxiliary engines have engine driven pumps with the lube oil system forming part of the engine itself.

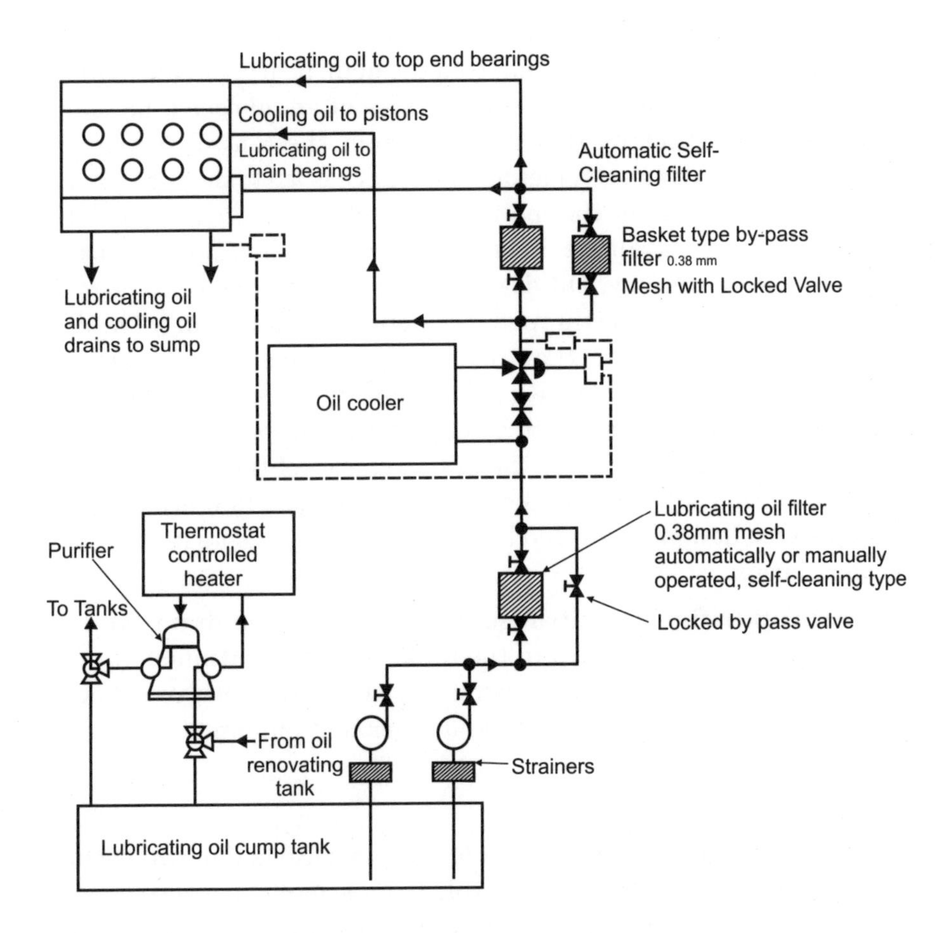

Fig. 4.16 Lubricating oil system.

Fig. 4.16 shows a typical lube oil system for a large two-stroke main engine. Circulating pumps draw oil from the engine drain tank, the suction pipe being sufficiently above the bottom of the tank to avoid sludge, water and other impurities from being picked up. The oil discharged by the pump passes through a filter which is either automatically or manually operating self-cleaning type, then through a seawater-cooled lube oil cooler. Oil pressure is always higher than seawater pressure to ensure that in the event of a leak seawater does not leak into the oil system. Temperature is controlled by an automatic mixing valve located at the oil outlet from the cooler that controls the oil flow rate through the cooler. It takes inputs from temperature sensors and maintains the oil temperature at the engine inlet. In case of an oil-cooled piston, temperate sensors at piston cooling oil outlets also transmit signals to the temperature controller. Next, the oil passes through a fine filter which is of the automatically self-cleaning type, then branches out to various distribution points that feed the main bearings, camshaft and chain drive, thrust bearing, piston (in case of an oil-cooled piston), to exhaust valve actuators, the crosshead and the control system that operates the reversing servomotor. One branch is led to the turbocharger in case it forms part of the main engine lube oil system. Circulating pumps, filters and strainers are in duplicate and when one is in use the other acts as a standby. Oil draining from the engine collects in the engine sump, then drains through strainers to the drain tank. The drain pipe is located remote from the circulating pump suction and submerged in oil in order to avoid aeration. The drain tank, located in the ship's double bottom, is provided with a vent pipe and a sounding pipe and is large enough to accommodate the full oil charge of the system. A sight glass located at each piston outlet reveals if adequate flow is maintained to all pistons. Low pressure and high temperature sensors in the system trigger alarms.

An **oil purifier** or centrifuge draws oil from the drain tank and delivers the purified oil back to the tank. It is run continuously at sea and when the engine is shut down batch purification is carried out as required to remove water, sludge and other contaminants.

Online Water Monitor

High water content in lube oil could lead to serious damage within a short period of time if not checked. It is important to strictly monitor water content in the oil and maintain it within acceptable levels as otherwise it can lead to corrosive wear and change the bearing geometry. By breaking the oil film within the bearing it can cause excessive wear and scuffing. Additionally, it can cause numerous other problems such as additive depletion, oil oxidation, corrosion, reduced lubricating film thickness, accelerated component wear rates, microbiological growth and loss of power. Therefore, continuous and accurate monitoring of water in oil is essential so that problems can be averted in time. Modern engines use the EASZ-1 system developed by Alfa Laval, which provides continuous online water monitoring to detect possible water contamination and trigger an immediate purifier desludging operation and also sound an alarm in the control room if the water ppm in oil exceeds a set limit, usually set between 0.2% and 0.5% water content. It measures the total water content in the oil whether dissolved, emulsified or free. When an alarm is sounded, water leakage into the lube oil system can be immediately investigated, e.g. from cooling water leaks, heat exchangers or purifiers. Sometimes if there is a fresh charge of oil in the system and the oil is free of other contaminants, the EASZ-1 is used for monitoring on-line the moisture content in lube oil and allowing oil purifiers to be started only when required. The unit responds very quickly to a change in the dielectric of the oil being monitored and starts up the purifier if moisture content exceeds a set limit.

Oil Maintenance

Lubricating oils are required to perform various functions in addition to lubricating and cooling and this places extensive demands on it during its working life. In a trunk piston engine for example, it is required to lubricate piston rings and liners, withstand very high temperatures and pressures and neutralize acidic products of combustion entering the crankcase due to piston blow past. In two-stroke engines the oil has to withstand high temperatures when used for piston cooling and withstand high mechanical loads in chain drives, gears and cams. The same oil is used as a hydraulic medium to operate exhaust valves and various servo-motors. It is therefore crucial to maintain the oil in a stable and optimum condition not only to ensure effective lubrication but also to arrest its degradation in use.

Before charging lube oil into a new engine, utmost care is taken to clean all engine components of metal particles, sand, dust and water. This is repeated on all the lube oil system components as well, like pumps, filters, pipelines, coolers and tanks (tanks are given a protective coating to prevent rusting). The entire system is then flushed with flushing oil till a clear discharge is obtained. Filters and strainers are cleaned at this stage before fresh lube oil is charged into the system.

During its working life, oil condition depends on how well the engine itself is maintained and operated. Oil can deteriorate due to oxidation brought about by the presence of oxygen, high temperature and the presence of catalysts. Oil is also vulnerable to contamination from areas like piston rings and liner, piston rods and stuffing box, scavenge spaces and water cooling passages in liner and piston crown. As stated earlier, water leaks especially cause not just deterioration of lube oil properties and its additives but also infect the oil with microbes. Moreover, water content above 1% has the potential to cause corrosion of bearing overlays leading to overheating and damage to bearings.

Microbial degradation occurs when organisms like bacteria, mould and yeast grow and multiply in the oil when water is present (from cooling water leaks or condensation in humid climate). They multiply fast in the further presence of carbon, nitrogen and sulphur contained in the oil and specifically in the temperature range between 20° C and 40° C. The bacteria deplete the oil not only of dissolved oxygen but also its sulphur content, leading to degradation of the oil and corrosion of steel.

Sliminess of the oil and foul smell are direct indications. Overgrown bacteria can be eliminated by use of fungicides or biocides followed by heating the oil to around 80° C while running the purifier continuously to clean and sterilize the entire oil charge. Subsequently, all tanks, pipes, filters and strainers should be thoroughly cleaned and maintained clean to prevent re-growth of bacteria. Tests include dipping a special gel in the oil and letting it incubate to develop a growth of bacteria. The colour of this culture is then compared to a standard chart to determine the degree of contamination.

Oil samples are taken at regular intervals from various points in the system for testing both on board and in a laboratory.

On-board tests include:

- Litmus test or colour titration by a portable unit to assess alkalinity
- Flash point test to find out possible fuel contamination.
- Viscosity test using a flow stick.
- Test for water in oil – a piece of paper dipped in oil will sputter when burned if water is present or, drops of oil on a hot plate will crackle if water is present or, using a hydrometer.
- Blotter test to indicate oxidation level and presence of suspended carbon.
- Microbial infection test using a special gel that cultures microbes.

Periodically, once in 1000-2000 running hours (for conventional engines), oil sample is taken from a running engine, from the same point, and sent ashore for laboratory tests. They include:

- Flash point test using standard equipments.(Pensky-Martins close flash point test)
- Spectro-analysis to determine contamination by metals like sodium, vanadium and chrome eroded from various engine parts, from combustion products of heavy oil and ingress of sea water.
- Density
- Water content by distillation method
- Tests to determine presence of insolubles like carbon, dirt and inorganic particles.

Apart from revealing the trend in the deterioration of the oil, like decreased alkalinity, anti-dispersants, antifoaming agents and antioxidants, and increased acidity, these tests also reveal the quantity and type of metal particles present in the oil. Because the linings of the different bearings in the engines are of different specifications, an analysis of the metal particles will readily reveal from which bearing the particles originated

In addition, visual observation too readily indicates various other contaminants in the oil. Oiliness can be found by rubbing oil between fingers. This will indicate contamination by fuel. Foggy sight glasses will indicate presence of water. Purifier sludge discharge will also show sediment accumulation and deterioration of oil.

Oil filters need to be cleaned at regular interval especially after a crankcase overhaul or drain tank cleaning. Wire-gauze elements should be cleaned thoroughly or replaced when needed. Care should be taken to avoid use of inflammable materials like spirit for cleaning. Pressure difference before and after a filter will readily indicate a dirty filter.

Lube oil coolers also need periodic cleaning, especially on the oil side, whenever pressure difference between inlet and outlet indicates a clogged or dirty cooler. Cleaning agents like coal-tar naphtha or trichloroethylene are generally used. The cooler is first drained of oil, the inlet and outlet valves shut and the oil side filled with the cleaning medium. After three to four hours the dirty fluid is drained into an empty drum – for reuse after due purification by on-board oil purifiers or sent ashore.

Cooling Systems

About 26% of the heat generated by the combustion of fuel in an engine is conducted to various engine components like piston, cylinder liner, cylinder head, exhaust valve and fuel injector, accounting for the biggest share of heat loss. Additionally, about 5 % of the indicated power generated by the engine is also lost in overcoming friction in the various rubbing and mating components like piston rings and liner, crosshead and guide shoes, piston rod and stuffing box, main and bottom end bearings, cam shaft etc. that accounts for further energy loss and heat generation. All this heat has to be carried away in order to maintain the engine at a safe operating temperature and to avoid thermal stresses. This is achieved by the engine cooling system that employs fresh water to cool the cylinder head, exhaust valve, injector, cylinder liner, piston, the turbocharger and the engine lube oil. The exceptions are piston and charge air. While the piston is oil-cooled in some engines, charge air is normally cooled using seawater. Smaller diesel engines may opt for distilled water as a coolant. However, though it has a higher specific heat and has no scale forming salts, distilled water is more expensive and if produced by evaporating seawater it would be also acidic.

The cooling system also facilitates gradual warming up of the engine prior to starting and controlled cooling down after shut down in order to avoid thermal shock. This is also important since large fluctuations in temperatures not only cause undue thermal stress, but can also cause rubber seals to leak.

Medium speed engines use fresh water circulated through cylinder jackets and cylinder heads for cooling. Even though this would be an independent system, there is usually a connection to the main engine jacket cooling water system, which is opened to circulate water to the jacket cooling water heater to warm up a cold main engine and to keep the engine warm during stand-by.

Additives

Since fresh water contains salts that can deposit on metal surfaces and cause corrosion, it is suitably treated to prevent scale formation and corrosion. The additives generally used are either anti-corrosion oils or inorganic inhibitors. Both build up a thin oily layer on metal surfaces that guard against corrosion. Chromates, sodium nitrate and emulsion oils are generally used. However, since chromate is toxic and harmful if it finds its way into drinking water, only the last two are approved for use where a fresh water generator is employed. These are non-toxic, safe and form a thin passive oxide layer on the metal. Though chemical additives like nitrate-borates do not harm rubber seal rings, they attack zinc. Therefore zinc anodes are avoided, as also galvanised pipes and soldered joints. Soluble emulsion oils offer better protection against cavitation and pitting. They form a greasy coating on metal surfaces which offers good protection against corrosion, but the oil concentration must be maintained within close limits, otherwise, there is a risk of narrow passages getting choked leading to overheating. Emulsion oil also deteriorates with time necessitating periodic system cleaning and annual recharging. Also, any biocide added to water to kill bio-organisms must be compatible with domestic fresh water generation if used.

Jacket Cooling System

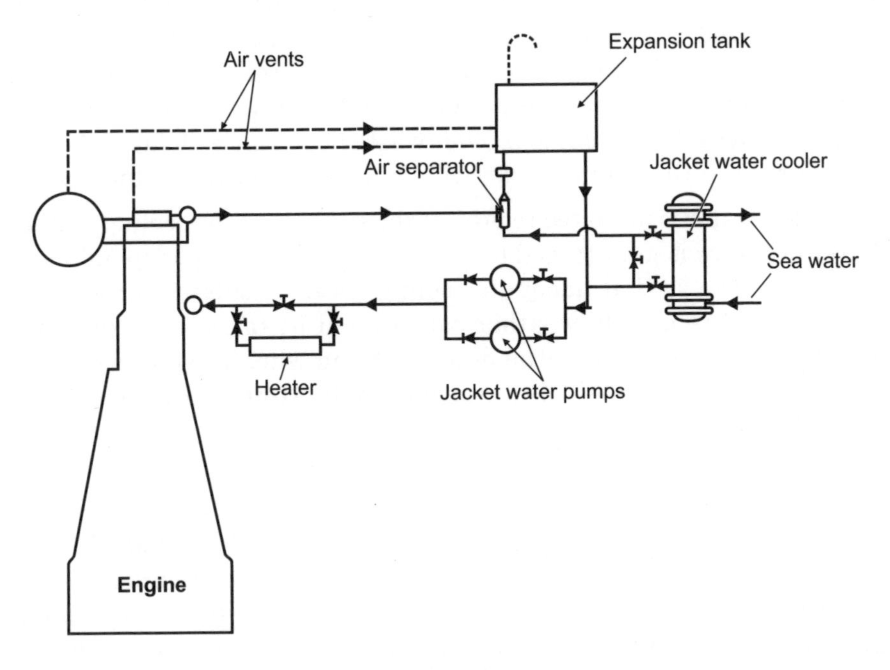

Fig. 4.17 (a) Jacket cooling water system.

Jacket water is inhibited to protect the surfaces of the cooling system against corrosion, corrosion fatigue, cavitation and scale formation. Fig. 4.17 (a) shows a schematic diagram of the jacket cooling water system for a two-stroke main engine. It forms a closed circuit with an expansion tank located well above the top of the engine for venting and pressurising the system. Heated water returning from the engine passes first through an air separator (to separate trapped air, since gas or air circulating in the system can reduce cooling considerably) and then through the jacket water cooler that is circulated with seawater. The cooled water is then led to the circulating pump suction, which delivers it to the engine supply manifold from where branch pipes lead the water to the base of individual cylinder jackets. A heater is incorporated between the pump delivery and the engine as an additional facility to warm the engine during start up. Exiting from the top of the jacket, the water then enters the cylinder head. If the engine has exhaust valves, the water circulates through these before finally joining the common collecting pipe. A branch pipe leads from here to the turbocharger, then returning the water back to the discharge line. The electrically driven circulating pump is in duplicate and the suction side is connected to the bottom of the expansion tank through an equalising pipe for effective venting and to maintain a positive pressure. Air vent cocks located at the highest points in the system, at the engine discharge line, turbocharger and cooler, also connect via rising pipes to the expansion tank for venting. An alarm device inserted between the de-aerating tank (air separator) and the expansion tank, signals if excess air or gas is released, indicating a malfunction in engine components. A fresh water generator is incorporated in the system to recover waste heat, since about 11 % of all the lost heat in the engine is carried away by the jacket cooling water.

Jacket water is inhibited to protect the surfaces of the cooling system. Water pressure is maintained between 1 to 1.5 kgf/sq cm depending on the engine size so that adequate cooling water circulates through the engine. In extreme cold weather care should be taken to ensure that liner temperature is above the dew point and when the engine is shut down cooling water should be drained from the system to avoid fracture due to freezing.

Water temperature at inlet to the main engine is maintained at 55° C at full load and five degrees higher at slow running. This is to ensure that the liner temperature is maintained between 150°C and 220°C, in order to minimise liner wear. Cooling water outlet temperature is maintained between 80-85°C. Temperature is regulated by controlling the seawater flow to the cooler (by throttling the by-pass) and is indicated by thermometers installed at all jacket outlet pipes and the turbocharger. Individual jacket temperatures can be adjusted by throttling the jacket outlet valve while all inlet valves are always kept fully open.

Automatic Temperature Control

While manual control of jacket cooling water temperature works well under steady load conditions, a more accurate control of the temperature is essential for efficient engine operation during fluctuating loads, especially during manoeuvring. This is achieved by incorporating an automatic temperature control system as shown in Fig. 4.17 b.

The system uses a combination of cascade and split range control. In the cascade control the output from a master controller is used to automatically obtain the desired value through a slave controller, both being identical. The master controller receives a signal from a temperature sensor in the engine cooling water outlet pipe. This is compared with a set value and any deviation is used to adjust and obtain the desired value of the slave controller. The slave controller also obtains a temperature signal from a sensor in the water inlet pipe to the engine, which it compares with its own set value and any deviation generates a signal that is transmitted to two control valves: one to control seawater inlet to a water cooler and the other to control steam inlet to a water heater, which forms the so called split range control. In the event the jacket cooling water temperature rises the seawater control valve opens to admit more seawater to the cooler and if the cooling water temperature drops, the seawater control valve will then be throttled. If however, the seawater control valve is fully closed and still the jacket water temperature is low, then the steam control valve will open to admit steam to the heater to heat the water.

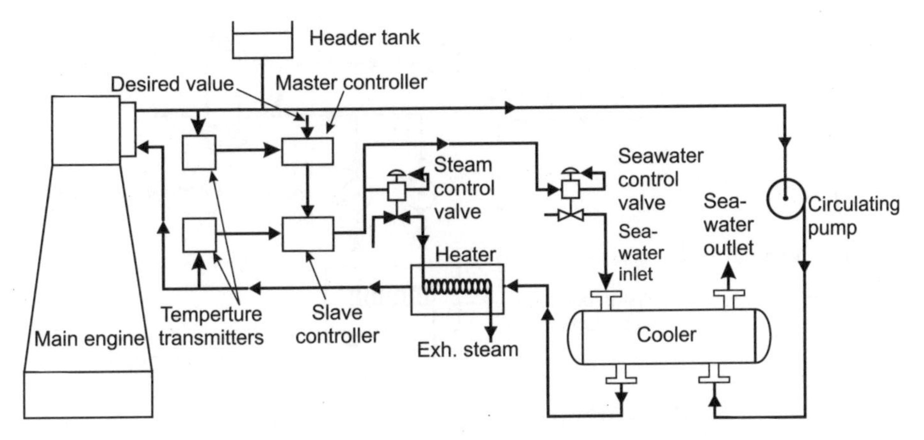

Fig. 4.17 (b) Jacket cooling water temperature control.

Maintenance

Leakages from the system should be kept to an absolute minimum. Otherwise it would not only mean constant replenishing of water in the expansion tank but also depletion of additives in the water, thus increasing deposits. Leakages normally occur from pump glands, pipe flanges, vent cocks and drain valves. However, even in the best maintained system, and despite adding additives, it is not possible to completely eliminate scale formation on the cooling surfaces. Therefore it becomes necessary to remove the scales periodically or when scale is observed during inspection of accessible spaces. This is done by circulating dilute hydraulic acid through the system, till the acid strength weakens after the scales are dissolved. The system should then be drained and flushed with fresh water. An alkaline soda solution should then be circulated to neutralize remnants of the acid and the system pressure tested to ensure that water seals are not leaking.

It is important to make sure that all vent cocks in the system are always kept open so as to allow air to escape. Accumulation of air in the cooling chambers and pipes could lead to fluctuations in the cooling water temperature and pressure. This could also occur if the expansion tank water level is too low or runs empty as this will reduce the static pressure head on the system. Another cause could be possible blockage in cooling water pipes.

Piston Cooling System

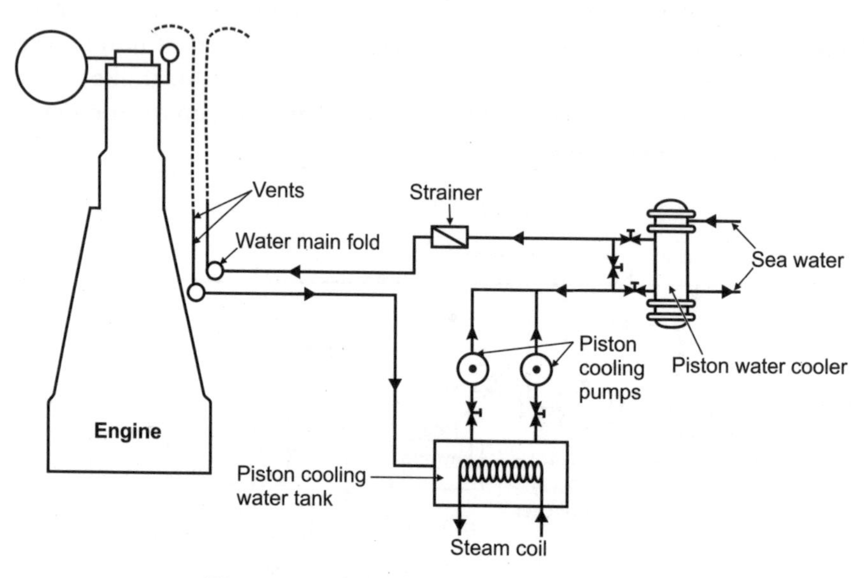

Fig. 4.18 (a) Piston cooling water system.

Since piston temperatures are higher than the cylinder liner, the piston is cooled by an independent system, which also prevents possible contamination of jacket cooling water should lube oil leak into the cooling water. Since the piston reciprocates up and down, some means of supplying the water to the inside of the crown must be available. Some engines use telescopic and stand pipe arrangements while others use lube oil as a coolant (supplied through a bore in the piston rod) although the cooling effect is not as good.

As shown in Fig. 4.18 (a) cooling water, after passing through the piston crown is allowed to flow out to a drain tank outside the engine. The circulating pump draws water from this tank and delivers it through a cooler to a manifold extending the length of the engine from which branch pipes lead the water to individual standpipes for each unit. The top ends of these pipes are sealed by a gland through which the inner telescopic pipes reciprocate as they deliver the cooling water to the inner recesses in the crowns. Returning water passes through a similar set of pipes then flows through a sight glass or flow indicator before joining up in an outlet manifold on its way to the drain tank. Air vents fitted to the top of the standpipe prevent water hammering – indicated by noise in the cooling system due to vapour lock. Leakage water from the glands collects in a chamber from where it is led outside the crankcase. Water temperature is controlled by throttling the cooler bypass valve and steam heating coils in the drain tank facilitate heating the cooling water to warm up the engine during start up.

Automatic Temperature Control

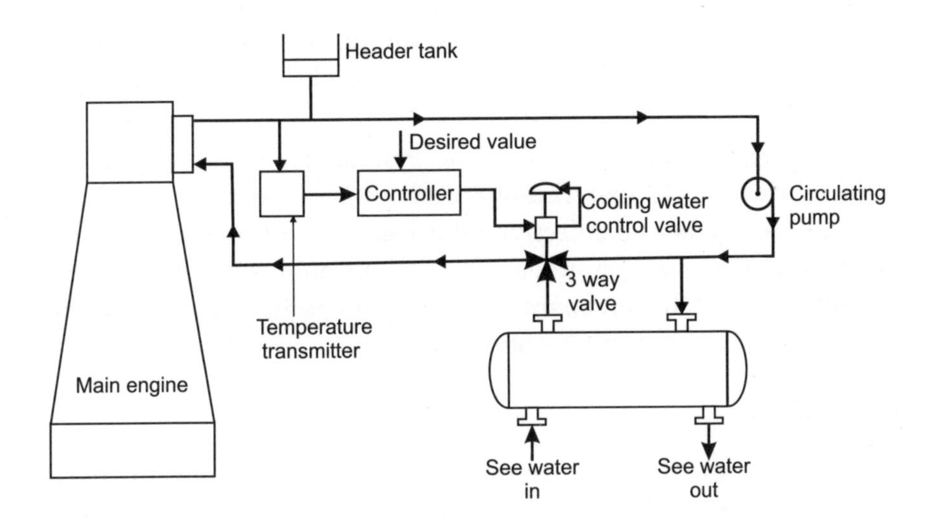

Fig. 4.18 (b) Piston cooling water temperator control.

For more accurate control of the piston cooling water, an automatic control system that uses a signal from a temperature sensor to actuate a three-way valve is employed (Fig. 4.18 b). Instead of a manual cooler bypass valve, a three-way control valve is employed to enable bypassing of the cooler. A constant full flow of seawater is maintained through the cooler. A sensor in the piston cooling water outlet provides a signal to a controller which compares it with a set value and any deviation between the two generates an output signal that actuates the three-way control valve. If the input signal indicates a low piston cooling water temperature, the bypass valve will open so that less water passes through the cooler thus rising the temperature and if the measured temperature is high the bypass valve will be throttled, circulating more water through the cooler and thus lowering its temperature.

Oil Cooling of Piston

Modern engines increasingly use oil for cooling the piston. Though specific heat of oil is lesser than that of water, resulting in poorer heat transfer, its advantages far outweigh this drawback: absence of scale formation, no contamination of the engine lube oil system, no corrosion and a simpler construction due to the absence of telescopic pipes. Additionally, the entire piston cooling water system consisting of pumps, pipelines etc could be avoided since piston cooling could form part of the engine lube oils system itself – necessitating only a larger system capacity. Cooling oil from the lube oil system is delivered through articulated pipes to the crosshead from where it rises up through concentric pipes in the bore of the piston rod to the crown. Returning oil is allowed to drop into the engine oil sump tank.

Some engines employ separate oil systems for piston cooling and engine lubrication. Though this offers advantages like the possibility of using different oils and selective additives for both the systems and improved control over piston temperature, it would nevertheless entail greater initial cost due to separate storage, pumps and pipelines and the care needed to prevent mixing of the two oils.

Maintenance

Poor coolant and faulty circulation will readily show in piston overheat – though this could also be due to engine overload, poor lubrication, faulty injector or piston rings, and insufficient scavenge air. However, if the fault is associated with the coolant, it can be ascertained by the following additional indications: fluctuations in the coolant pressure and temperature and deposits of scale or carbon in the cooling passages observed through inspection windows. The remedy lies in ensuring vent cocks are open, cleaning the heat exchanger, checking the circulating pump for proper delivery, checking coolant pipes for obstruction and, if indicated, de-scaling the system. Running checks include keeping a watch on flow indicators in cooling oil return lines, maintaining proper engine bearing clearance so that oil does not escape too freely depriving the piston of enough cooling oil and ensuring cooling oil outlet temperature does not exceed 70°C. Periodic maintenance include checking reciprocating telescopic pipes and glands every 1000 hours and replacing gland rings if necessary and taking out the piston once a year for inspection and checking and cleaning all cooling spaces.

Fuel Injector Cooling

This is a smaller system compared to jacket and piston cooling and forms a closed loop with its own expansion tank. Cooling water is circulated through drilled passages in the nozzle to carry away the intense heat of combustion to which it is exposed. Jacket cooling water is used as the cooling medium to cool injector cooling water. Jacket cooler outlet line is tapped and a branch line supplies cooled jacket water to the injector cooler before draining into the jacket water drain tank. Injector cooling water temperature at the nozzle outlet is maintained at about 92° C to prevent it from flashing into steam. In order to maintain the temperature within close limits automatic control systems are sometimes installed.

Seawater Circulation System

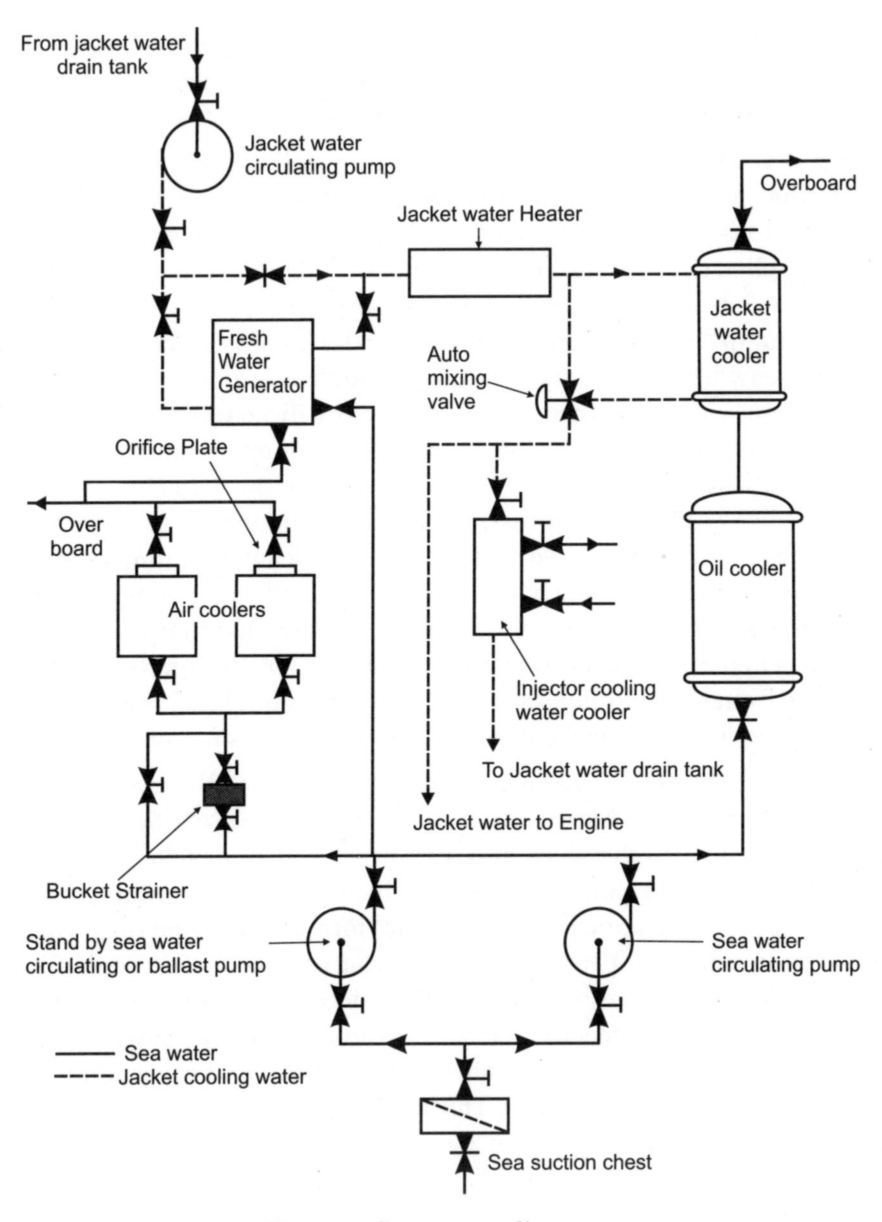

Fig. 4.19 Seawater cooling system.

Seawater is used as the cooling medium in the following equipments:

- Jacket cooling water heat exchanger
- Piston cooling water heat exchanger
- Lube oil cooler
- Charge air cooler

In addition, seawater is used for evaporation into fresh water in fresh water generators. Seawater also flows through condensing coils in the generator to condense the steam and works the air ejectors to create the necessary vacuum required for evaporation.

Fig. 4.19 shows schematically a seawater cooling system. Seawater circulating pump (in duplicate) draws water from a sea suction line connected to sea suction valves or 'sea chests', one on each shipside. The pump supplies the piston cooler and the lube oil cooler, the water then flows through the jacket cooler before discharging overboard. One branch from the pump delivery supplies the fresh water generator while another provides seawater to charge air coolers through a strainer. Both join up to discharge overboard through an overboard discharge valve. A sea strainer before the pump suction protects the pump from extraneous matter. Shipside sea chests are located well below the waterline so that there is always a positive pressure at pump suction. A drop in this pressure could mean a clogged suction strainer. Seawater pumps supply cooling water to the auxiliary engines as well and to air compressors. However, a smaller seawater pump is always provided that runs when the ship is in port and sea water is required only for the auxiliaries. The discharge from the auxiliary engines could be re-circulated through the jacket water cooler for preheating of main engine when required.

To avoid corrosion, seawater pipes are generally made of an aluminium-brass alloy. Further protection is provided by sacrificial anodes fitted inside heat exchangers and strainers. A new development has been the use of a titanium alloy to make heat exchangers, piping etc. A stable, inert and permanent oxide film that forms over the metal surface makes titanium almost immune to seawater corrosion. The oxide film has been found to effectively prevent erosion, cavitation and pitting. Titanium alloys are especially resistant to erosion so that seawater flow rates of 30 m/s can be safely handled even if sand and other abrasive particles are present in the water. This material also withstands oxidising and reducing conditions created by microbiological organisms in the water.

Central Cooling System

The central cooling system, Fig. 4.20, employed in some MAN B&W engines is an alternative to seawater cooling system. Seawater is supplied only to a central cooler, and then discharged overboard. Inhibited fresh water circulated through this cooler by a set of central circulating pumps supply the cooling medium to the various main engine heat exchangers and also to the auxiliary engine coolers. Freshwater at a lower temperature is circulated by the central cooling water pump through a cooling circuit passing through the main engine lubricating oil coolers, the scavenge air coolers and the auxiliary engines, which have their own engine-driven pumps for circulating jacket cooling water. A separate independent main engine jacket cooling water pump circulates the cooling water through the main engine and then to the fresh water generator and the jacket water cooler. The cooled water then passes through the de-aerator and joins the jack water pump suction side to continue the circulation. A thermostatically controlled 3-way valve at the jacket cooler

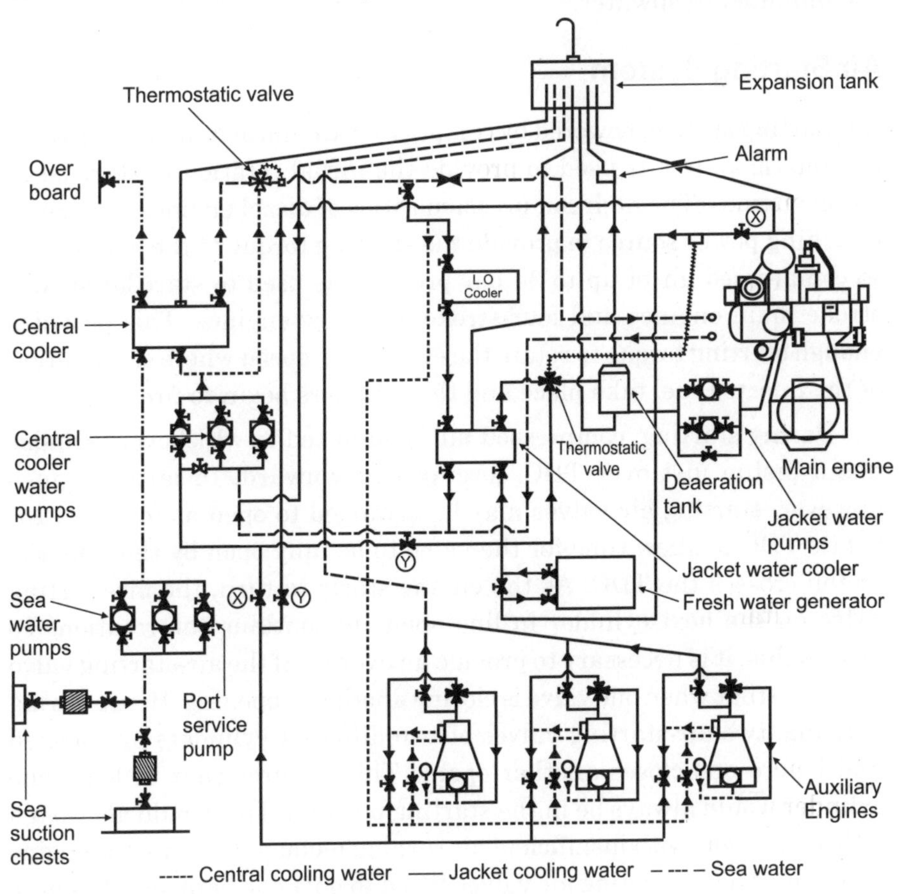

Fig. 4.20 Central cooling system.

outlet mixes cooled and uncooled water to maintain a main engine outlet water temperature of 80 to 85 °C. During port stay, in order to warm up the main engine, valves X are closed and valves Y open. The small stand-by central water pump circulates the necessary flow of water for the air cooler, the lubricating oil cooler, and the jacket cooler of the auxiliary engines. The auxiliary engine-driven pumps, circulating cooling water through the interconnecting loop with the main engine system, ensure a satisfactory jacket cooling water temperature of the auxiliary engines. The loop incorporates a by-pass system integrated in the main engine jacket cooling circuit allowing pre-heating of the main engine.

There is only one expansion tank and to prevent the accumulation of air in the system a de-aerating tank is located below the expansion tank with an alarm device incorporated between the two to give a warning when there is excessive air release. Maintenance work is minimised by the use of a central cooler, as this is the only component that is in direct contact with seawater and all other parts of the system use inhibited freshwater.

Air Starting System

Air starting system, powered by compressed air and actuated by a hydro-pneumatic system is used to provide the initial rotation to start large diesel engines. The high compression ratios of diesel engines requires a matching power source to provide the starting torque to the engine and so compressed air of up to 30 bar pressure is used to start large two-stroke main engines and four-stroke auxiliary engines. This provides enough starting torque to rotate the engine to a speed where self-ignition of the injected fuel take place and the cylinders begin to fire.

During starting, compressed air is admitted to whichever cylinder has a piston just over TDC, forcing it downward. In large engines however, starting air valves may be arranged to open as much as 10° before TDC to allow time for the valve to be fully open by the time the piston crosses the TDC. As the engine starts to turn, the air-starting valve on the next cylinder in line opens to continue the rotation. To ensure this, it is necessary to provide an overlap of the air-starting valve timing so that when one valve is closing another is opening. Hence during starting, two air-starting valves of two different cylinders are opened and before one closes another opens. This ensures that at least one cylinder would always be in the starting position. This would also avoid a 'dead position' with insufficient air turning moment to rotate the engine (V-engines have starting air valves fitted to only one bank of cylinders).

The minimum overlap provided is 15° (ideal would be between 20° to 90°). The valve must also close at least 5° before exhaust valve opens in order to prevent high pressure starting air from being blown through the exhaust (also because exhaust manifold is not designed to withstand a pressure of 30 bar). All this is accomplished by the control air system. Once the engine picks up the required speed (8 to 12% of the maximum rated speed), fuel is injected into the cylinders, combustion occurs, and the starting air is cut off. Initially some cylinders may misfire

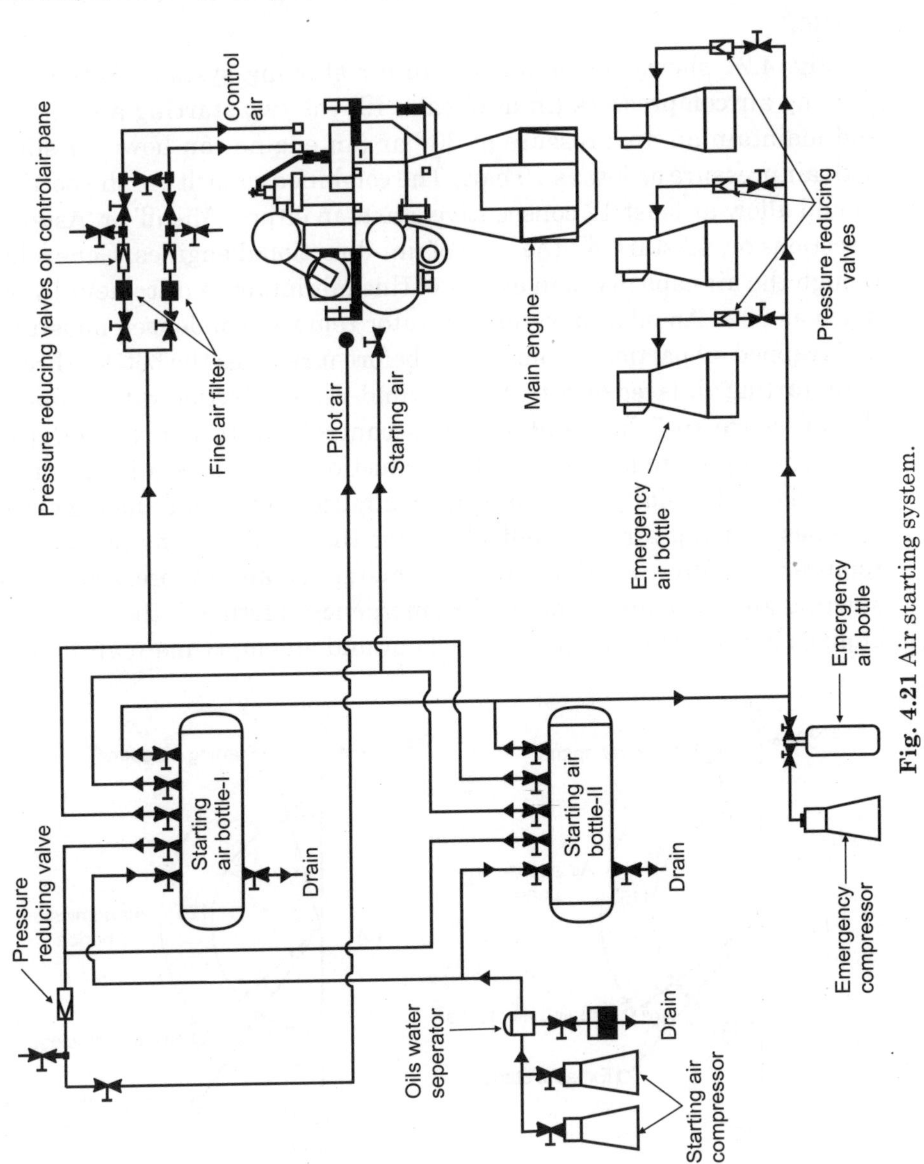

Fig. 4.21 Air starting system.

and the engine speed may pick up in jerks. It takes anywhere between 2 to 8 seconds after air-cranking for the cylinders to fire regularly and the engine to run smoothly under its own power.

Before this however, a large engine is first "blown through" with fuel set to zero and the indicator valves open, to clear the engine of any water build up and to ascertain that all moving parts are free. After blowing through both in 'Ahead' and 'Astern' directions, the indicator valves are closed on all cylinders and the engine is then ready for starting.

Fig. 4.21 shows schematically an air starting system. Automatic starting air compressors (in duplicate) fill the two starting air bottles and maintain an air pressure of 30 bar (an engine can however start with air pressure as low as 10 bar.) The combined capacity of the bottles should allow at least 12 consecutive starts in either 'Ahead' or 'Astern' directions or, 12 starts in the case of uni-directional engines. Generally though the air capacity is in excess of this minimum requirement by as much as 50%. An oil and water separator removes condensed moisture and trapped oil particles from the air before it reaches the bottles. From here starting air is fed directly to the cylinders. A reduction valve reduces the pressure to 7 bar and supplies control air to run the engine manoeuvring system. The pressure reduction however would result in high relative humidity necessitating an air drier. Another reducing valve supplies starting air and control air for the auxiliary engines at the required pressure. Additionally, an emergency air compressor and a starting air bottle are installed for emergency starting of the auxiliary engines in the event of a power blackout and an empty main air bottle.

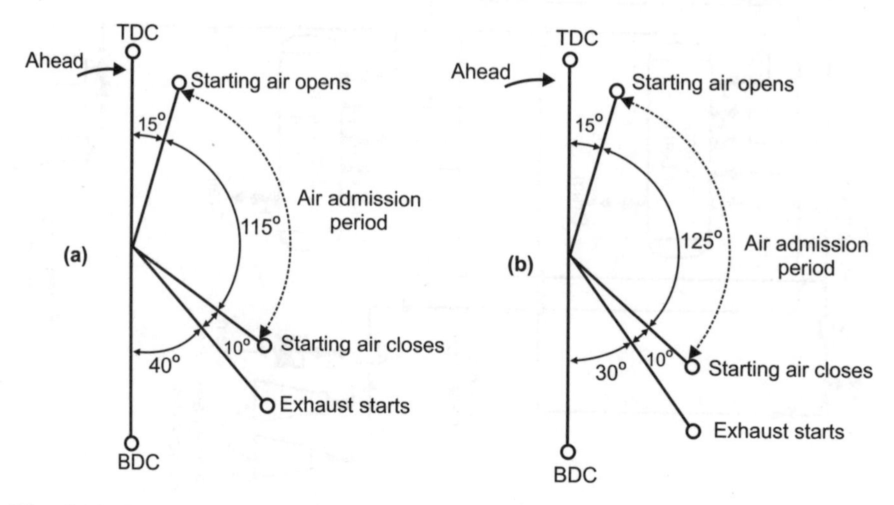

Fig. 4.22 Starting air timing diagrams. (a) Two-stroke cycle (b) Four-stroke cycle.

Consider Fig. 4.22 (a) for a 4-cylinder 2-stroke engine. The starting air valve opens 15° after TDC and remains open till 130° after the dead centre, giving an air admission period of 115°. The angle between cranks or the firing interval (number of degrees the engine turns before firing begins) for a 4-cylinder engine would be 360/4 = 90°. This would give an overlap of 25° (115–90), which would crank the engine quite satisfactorily in the firing sequence of 1 4 3 2. This firing order would prevent two adjacent cylinders from firing and thus prevent excessive loading on the crankshaft journal and also allowing better and regular crankshaft rotation. Firing sequence is thus related to engine balancing.

Similarly, referring to Fig. 4.22 (b) for a 7-cylinder 4-stroke engine, the starting air valve opens 15° after TDC and closes 140° after the dead centre, giving an air start period of 125°. The firing interval would be 720/7=102.85°. This provides an overlap of 22.15° (125–102.85) which is quite sufficient to crank the engine in this sequence: 1 7 2 5 4 3 6. From this it can be seen how valve overlap is related to number of cylinders. This is a crucial consideration since an overlap of less than 15° will not provide sufficient turning moment required to crank the engine. Overlap also depends on the air admission period and the exhaust timing, since starting air valve must close before exhaust commences.

System Components

Operating controls of the air starting system include the hand wheel of the automatic starting air shut off valve, the engine room reply telegraph, the starting lever and the fuel control lever at the control stand. Various other components that make up the starting and manoeuvring system are described below.

Automatic Starting Air Shut Off Valve

It is fitted to the starting air pipeline connected to the engine and comes into operation only during manoeuvring, remaining automatically closed when the engine runs on fuel (for extra safety it is manually closed on completion of manoeuvring). The valve has multiple functions: it acts as a stop valve supplying starting air from the air bottles to the starting air manifold and from there to individual cylinder air-starting valves during the starting manoeuvre and also shuts off the air when the engine runs on fuel. It acts as a non-return valve preventing blow back of combustion gases in the event of a leaking air-starting valve. Further, in case of a fire in the air-starting pipeline, it also prevents propagation of a flame on account of the flame trap incorporated in it.

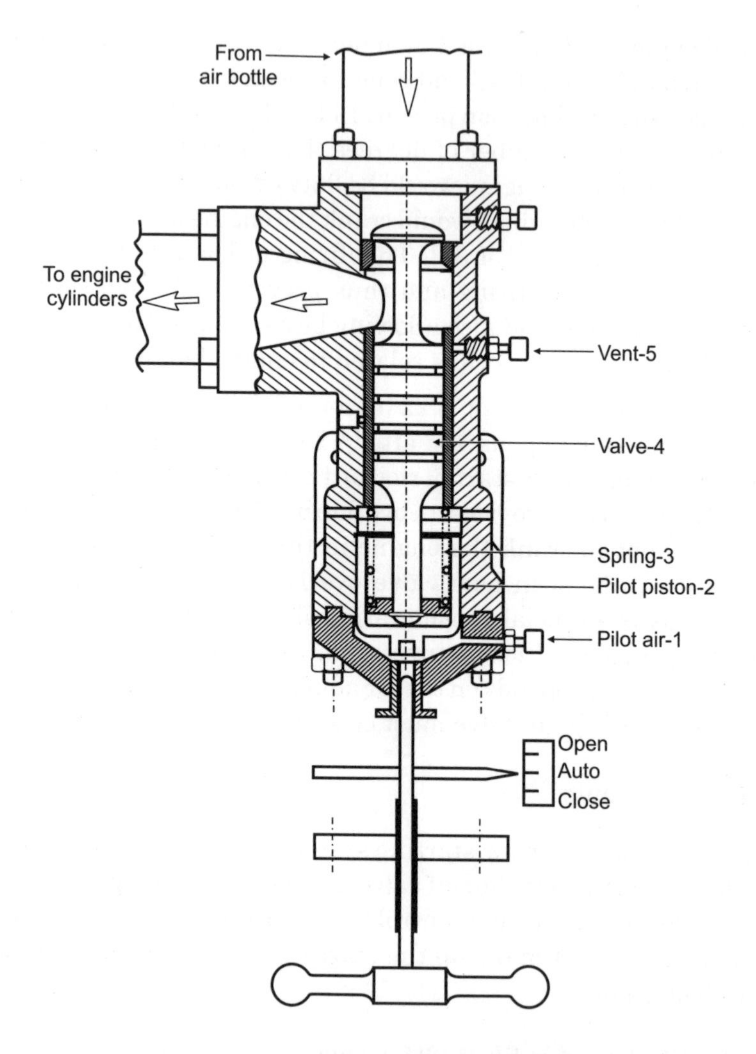

Fig. 4.23 Automatic Starting Air Shut Off Valve.

Fig. 4.23 shows a balanced type valve operating on positive pilot air pressure (as opposed to an unbalanced type that works when the pilot air pressure is relieved). Pilot air released from the pilot air valve reaches the connection 1 and acts on the pilot piston 2 pushing it up against the combined pressure of spring 3 and air pressure acting on the valve from above. This opens the valve 4 and starting air passes across valve seat 7 and reaches the air-starting valves. As soon as the cylinders fire and the engine begins to run on fuel, the pilot valve closes, shutting off air supply to the pilot piston and closing the valve swiftly under the combined spring load and air pressure acting on the spindle from above. Air in the starting air manifold is then vented through the connection 5.

Start Air Distributor

Fig. 4.24 shows one starting control valve inside the air distributor based on a Sulzer design. Similar starting control valves for other cylinders are mounted radially around the distributor cam such that all follower rollers bear against the cam. The distributor controls the opening and

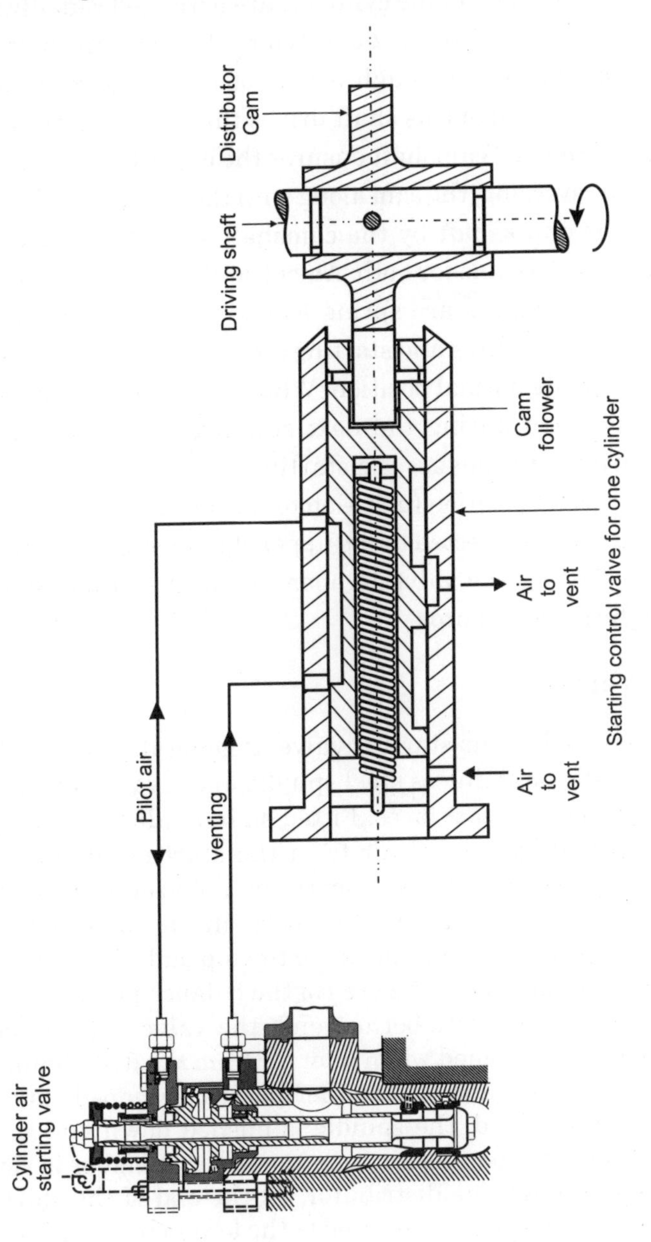

Fig. 4.24 Start Air Distributor.

closing of the cylinder air-starting valves in the proper sequence for starting or reversing the engine, by supplying or cutting off pilot air to the pilot piston attached to the valves. The distributor also helps in venting the lower chamber of the air-starting valves once air is cut off to a valve. As seen in Fig. 4.23, individual starting-control valves with their rollers, one for each engine cylinder, are arranged radially around a cam driven by a vertical shaft connected to the fuel camshaft next to the reversing servomotor, through a pair of helical gears (a gearbox coupled to the upper end of this shaft drives the engine governor). This direct drive from the fuel camshaft ensures the correct timing and firing sequence. While reversing, the cam along with the driving shaft is turned in relation to the crankshaft by the camshaft servomotor. The timing therefore remains the same for both 'Ahead' and 'Astern' rotation. Since the starting-control valves are spring loaded, they retract away from the cam once the engine fires and starting air is cut off; when the starting lever is returned to its normal position. When the starting lever is pulled to the 'Start' position during the next starting manoeuvre, pilot air reaching the distributor pushes the starting-control valve of a cylinder which is in the correct position to fire, on to the depression in the cam. In this position pilot air is directed to the upper chamber of the cylinder air-starting valve causing the valve to open, while the lower chamber is vented through the distributor.

Air-starting Valve

Fig. 4.25 shows a Sulzer air-starting valve. It generally has a mild steel body, a high tensile or stainless steel spindle and a valve and seat with stellited or hardened contact faces. Fitted on individual cylinder heads the valve is opened by control air from the starting air distributor. Starting air at a pressure of 30 bar from the manifold enters the chamber above the valve via the circumferential ports in the valve body. Under the upward pull of the spring attached to the top end of the spindle, the valve remains shut; and since the area of the balance piston is the same as that of the valve lid at its bottom end, the valve is pneumatically balanced. The valve is opened when pilot air from the distributor enters the top of the valve body and acts on the operating stepped piston. This overcomes the spring load, the spindle is pushed down and the valve opens. While this is happening air from the lower chamber is vented to the atmosphere through the distributor. At the end of the starting air admission period control air is admitted to the lower chamber to close the valve under the combined force of the spring and the air pressure acting

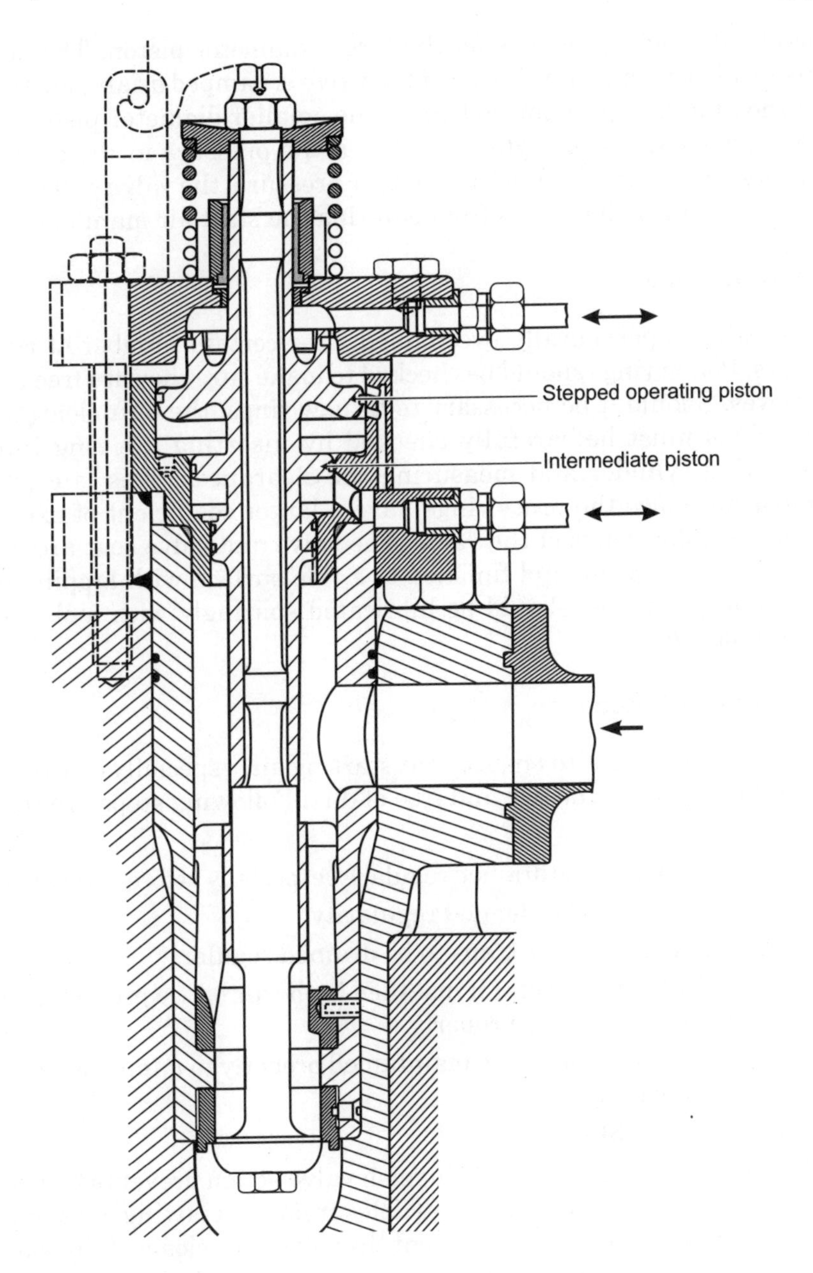

Fig. 4.25 Air-starting Valve.

upward on the intermediate valve. The upper chamber is meanwhile vented. At the end of the starting sequence air pressure is vented through holes in the main start air manifold.

Air on both sides of the operating piston maintains positive closing, and since it is stepped, once the valve begins to open, the opening is accelerated when air acts on the larger diameter piston. The stepped piston also means that closing of the valve is damped as air gets trapped in the annular space formed when the smaller diameter piston enters the upper part of the cylinder. Also, if the pressure in the cylinder is substantially higher than the start air pressure, the valve will not open, thus preventing hot gases from entering the start air manifold.

Maintenance

The valve is periodically overhauled after certain number of running hours. Piston rings should be checked to make sure they are free in their grooves. Should it be necessary to fit new rings, the butt clearances of the rings must be carefully checked by inserting the ring into the operating cylinder and measuring the clearances. This is especially important since the brass rings have a larger coefficient of expansion than the other parts of the valve. The valve and valve seat are ground with grinding paste and finished to a fine surface with lapping paste. All sliding surfaces should be lubricated sparingly with molybdenum disulphide grease.

General Maintenance

It is very important to ensure that starting air, especially control air, is maintained clean and dry and free of dirt. Following steps are taken to ensure this:

- Air bottles to be drained regularly, especially before manoeuvering.
- Air filters to be cleaned regularity
- Automatic water traps to be drained regularly
- Oil/water separator if installed in the air bottle pressurising line to be maintained properly.
- Air compressors to be maintained properly to minimise carry over of oil with air.

Running Troubles

The valve can begin to leak if sluggish valve action prevents fast closure of the valve, or dirt or foreign particles from the starting air supply get lodged on the valve seat and prevent the valve from closing fully. Sluggish valve action may be caused by dirty pistons or valve spindle guides or by parts fitted with inadequate clearances.

A leaking valve is indicated by overheating of the branch pipe connecting the starting air valve to the starting air manifold. During running, this should be checked regularly since a leaking starting-air valve can lead to an explosion in the pipe line. If a valve remains jammed open, fuel to the affected unit should be cut off. Since one unit would be out of operation load on the engine should be kept to a minimum and as soon as safe to do so, the engine should be stopped and the defective valve replaced.

Running Direction Safety Interlock

Fitted at the forward end of the fuel camshaft, its function is to block fuel supply during manoeuvring as long as the running direction of the engine does not coincide with the direction set on the engine room reply telegraph. The interlock will therefore allow fuel to the engine only if the direction of engine rotation agrees with the telegraph position.

Oil and Water Pressure Safety Cutouts

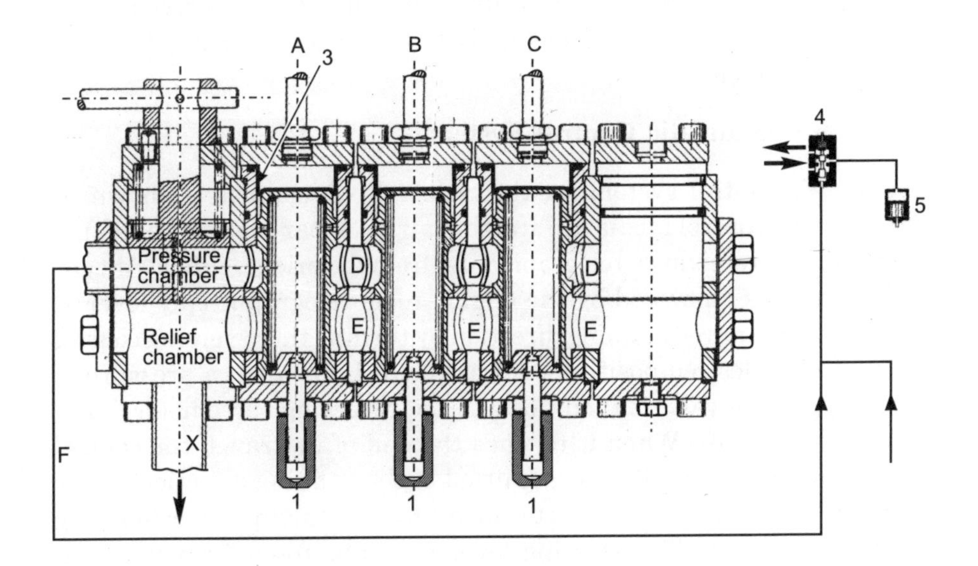

Fig. 4.26 Oil and Water Pressure Safety Cutouts.

This cut out is meant to safeguard the engine against possible piston seizure or overheating of piston or liner in the event of failure of the lube oil pressure or piston or jacket cooling water supplies. As shown in Fig. 4.26, normal working pressures of the afore mentioned three mediums are set on the cut out by turning the screws 1 as required to create the appropriate spring force under the pistons 2 that would force the piston up. This is countered by the downward force exerted by the three working mediums acting on top of the piston through connections A, B and C, thus keeping the piston under balance. In the event of drop in pressure in any of the working mediums, the corresponding piston will raise, pushed by the spring force from below, connecting spaces D and E across edge G.This will relieve the control oil pressure in the chamber D into chamber E below. This in turn will relieve the pressure in the control line F. The consequent drop in pressure through control valve 4 will cause the cut out servomotor 5 to move into the stop position, thus cutting off fuel supply from fuel pump to injectors. The top face of each piston is sealed against the working mediums by a diaphragm 3. It is possible to override the cut out in an emergency by closing the cock 6 so that pressure in the control line F is prevented from dropping, thus enabling the engine to continue running, as far as possible at a much reduced speed.

In the event the cut out is triggered due to drop in pressure of any of the mediums to a critical level, the fuel control lever must be immediately brought to a position of 3.5, so that if the pressure drop is corrected and the cut out device restores fuel supply to the injectors, heavy firing of the engine is avoided.

The Hydro-pneumatic Control System

Starting air control system for the Sulzer RND engines is shown in Fig. 4.27. The control media are starting air at 30 bar pressure and lube oil at 6 bar. On receiving a telegraph order from the bridge, either "Ahead" or 'Astern', it is answered by the engine room telegraph reply lever 1. This sets the required running direction by turning the reversing control valve 2 to the desired position via linkage A. Lube oil at a pressure of 6 bar supplied to valve 2 then flows to the reversing servomotor 3 and turns the camshaft. When it reaches the end of its travel, the control valve allows pressure oil to be admitted to the running direction safety interlock 4, which in turn directs the oil to the starting lever blocking device 5 via line B. The starting lever 6 is thus freed for movement. At the same time, pressure oil travels through line C to the slide valve 7, pushing the slide up and allowing pressure oil to travel to the fuel cut-out servomotor 8 which frees the fuel control linkages D to take a position corresponding to the position of the load indicator 9 which, in turn, is determined by the position of the fuel control lever 10 (set to a position, around 3.5, sufficient for the engine to fire and pick up speed on air cranking). All these sequence of movements ensure that all working pressures of the engine (lube oil and that of the jacket and piston cooling systems) are above the minimum set on the safety cutout device 11. In the event that this condition is not satisfied and any working pressure is below normal, the slide in the valve 7 will move downward instead, and relieve the pressure on the piston of the fuel cut out servomotor 8 moving the fuel linkages in the opposite direction bringing it to zero position. The fuel control lever thus gets locked at zero on the

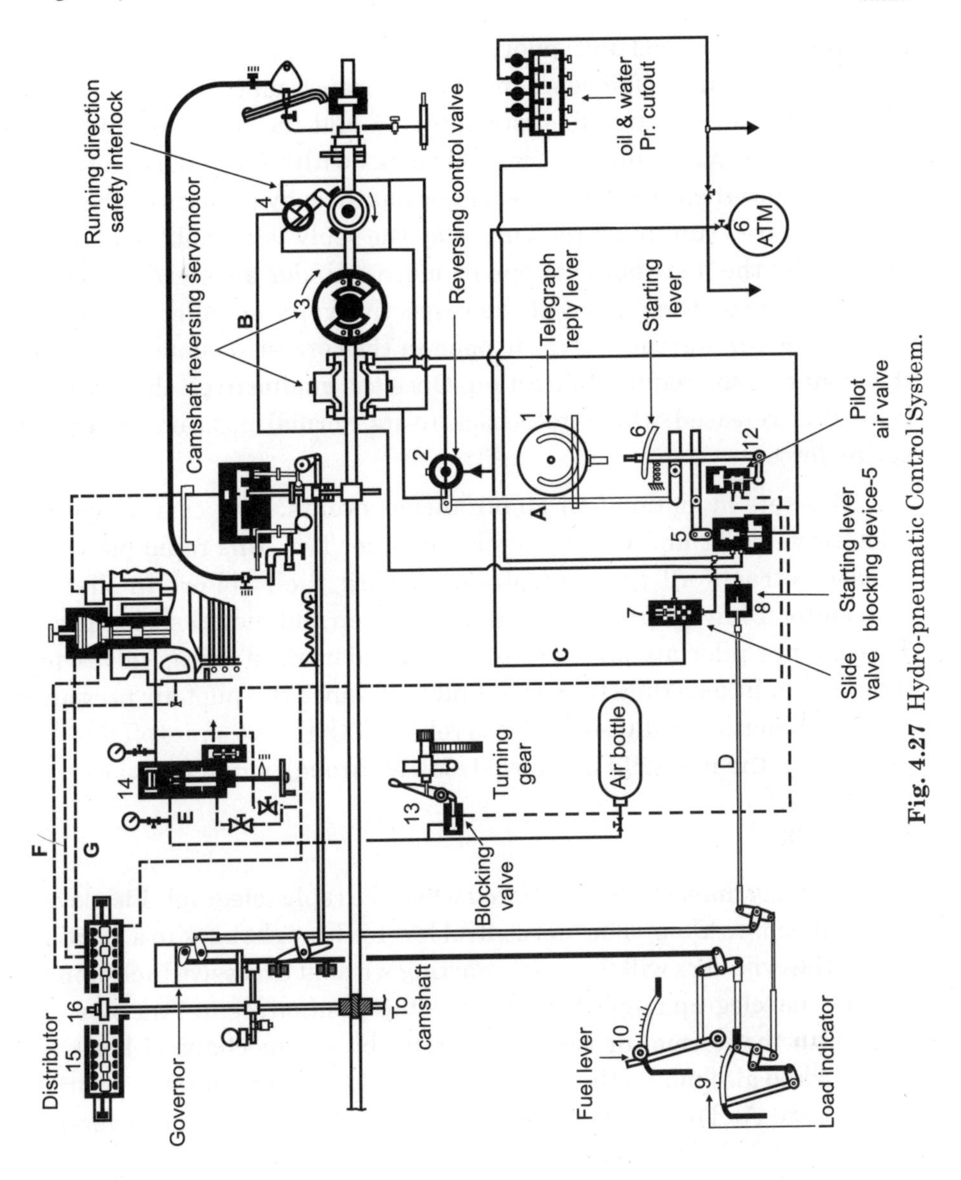

Fig. 4.27 Hydro-pneumatic Control System.

fuel rack (in an emergency though, it is possible to override the safety cut-out and move the fuel lever).

To initiate starting, the starting lever 6 is pulled to 'Start' position. This movement raises the pilot air valve 12, allowing pilot air to travel through the turning gear interlock 13 (provided the turning gear is disengaged) to the automatic starting air shut-off valve 14 through line E. The valve is thus opened and admits starting air to the engine manifold and thence to air-starting valves on individual cylinders, which would however remain closed until pilot air is admitted to actuate them. This is brought about in the following manner: a branch pipe from line E also connects to the starting air distributor 15 and the pilot air admitted through this forces individual control valves with their rollers on to the cam 16. Since the reversing servomotor has already positioned the cam to the required running direction set by the reply telegraph, pilot air is directed by the distributor to one or more cylinder air-starting valves which, acting on the pistons of the respective valve spindles via line F, actuate the air-starting valves to open in the correct sequence to crank the engine in the required direction. Once the engine fires, the starting lever 6 is released which goes back to its normal position under its spring force.

This sets in motion four simultaneous events. One, this closes the pilot air valve 12 and vents the pilot air line. Two, this rapid pressure release retracts all the control valves with their rollers in the air distributor 15 away from the cam, under their individual spring force. Three, since pilot air pressure to the cylinder air-starting valves too collapse via line G, they too close. And four, since the pilot air pressure to the automatic air shut-off valve is relieved, the valve shuts off too. Air pressure in the manifold is relieved slowly through small leakages.

Reversing

The reversing manoeuvre is initiated when the reply telegraph 1 is moved from “Ahead” to ‘Stop’. The fuel control lever is brought back to a position where the cylinders will fire on re-starting without excessive fuel supply. Moving the telegraph reply lever to the ‘Stop’ position secures the starting lever 6 in the normal or resting position by the mechanical blocking device 5 and also moves the reversing control valve 2 to the stop position via linkage A. This relieves the oil pressure acting on the reversing servomotor 3. This collapse of pressure makes the slide valve 7 move down under its spring force which in turn relieves the pressure on the piston of the fuel cut-out servomotor 8 which then cuts off fuel injection. The downward movement of the slide valve 7 also relieves oil pressure from the hydraulic component of the blocking device 7 thereby blocking the starting lever hydraulically too. Once the engine speed drops sufficiently, the reply telegraph is moved to ‘Astern’, which moves the reversing control valve 2 to the astern position. This makes it convey pressure oil to the reversing servomotor 3 thereby turning the camshaft to astern position. On the camshaft reaching the end of its travel, the running direction interlock 4 will allow oil to travel to the hydraulic blocking device 5 which will in turn free the starting lever. The mechanical component of the blocking device would have already been released when the reply telegraph was moved from 'Stop' to the running position 'Astern'.

Next, on pulling the starting lever to the 'Starting' position, all the sequence of events described above for starting the engine in the 'Ahead' direction get repeated. Once the engine turns astern, and the running direction corresponds to the position of the reply telegraph, the running direction safety interlock 4 releases fuel.

In an emergency, the reply telegraph can be brought from the "Ahead" position straight to 'Astern' without pausing at 'Stop' for engine speed to drop. The moment the reversing servomotor completes its travel to the astern position, the hydraulic blocking device will release the starting lever and the engine can be started immediately. The forward running engine stops quickly and starts in the astern direction, and as soon as this happens, the running direction safety interlock 4 releases fuel. The fuel control lever may have to be moved to a slightly higher position than 3.5 to assist quick restarting of the engine. It is to be understood that this manoeuvre puts the engine under high stresses. Owing to the high momentum of a moving, loaded ship the drift will make the propeller turn in the ahead direction and when the engine is immediately run astern a high torque is developed to overcome the kinetic energy of forward motion. This manoeuvre therefore should be resorted to only in case of an emergency.

Electro-pneumatic Control System

Modern technology has largely replaced the hydraulic control systems used in earlier conventional engines. The starting and manoeuvring system of MAN - B&W SMC engines is explained below.

The engine can be operated from the engine control room or the bridge – through a remote control system that accomplishes the following operations: setting direction of rotation, starting the engine, speed control and stopping the engine. In the event of failure of the remote control system, engine can be manoeuvred from the engine control room.

The direction of rotation is set by adjusting the timing of starting air distributor and the fuel pump. Referring to Fig. 4.28(a), as soon as the starting lever is moved to say, 'Ahead' direction, at the control console in the engine control room or the bridge, a signal from the console reaches the remote control valve which releases control air through respective solenoid valve to the starting air distributor where, through a link system operated by a piston and cylinder, the distributor cam is rotated relative

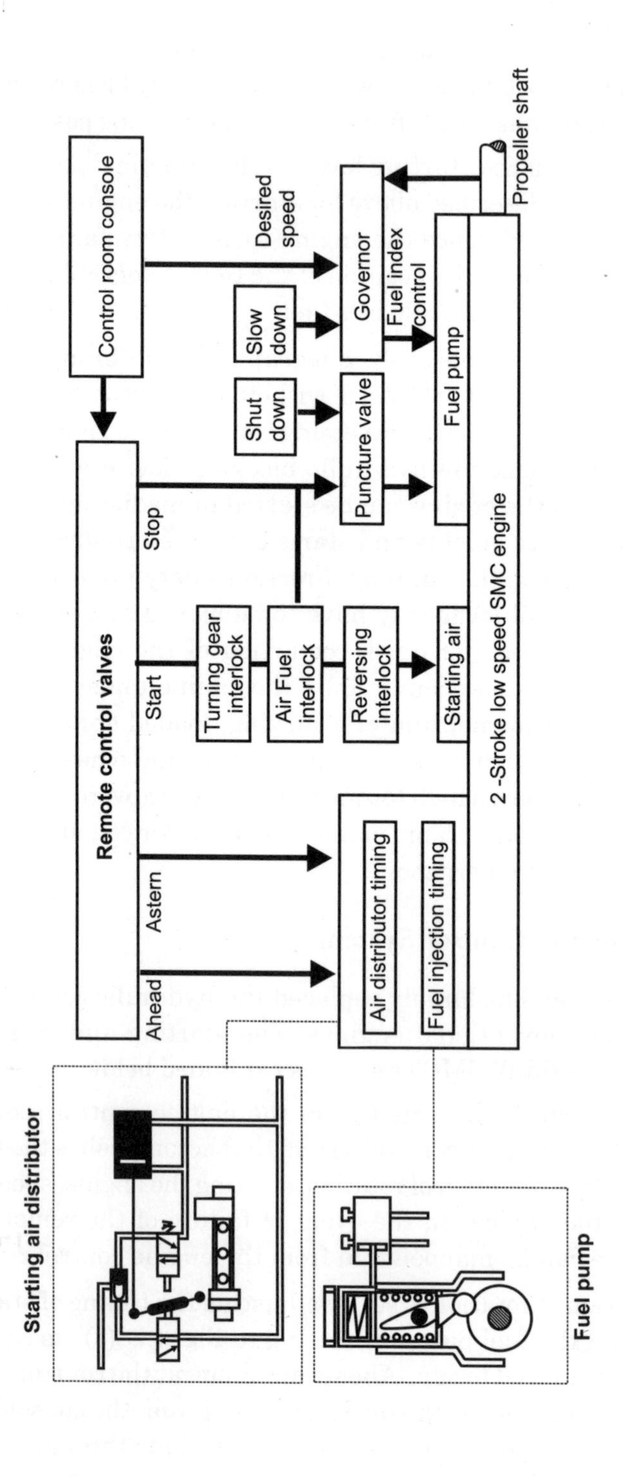

Fig. 4.28 (a) Schematic diagram of the electro-pneumatic control system of MAN - B&W engine.

to the crankshaft to start the engine in the desired direction. At the completion of this movement of the cam, a signal is sent to the reversing interlock to release a block on the fuel pump, which would normally prevent injection of fuel until the reversing sequence is completed. On release of this block, another control air signal is sent to the reversing mechanism of the fuel pump. This moves a link, actuated by a reversing cylinder that displaces each cam follower relative to the cam, to alter the pump timing. The timing diagram for both 'Ahead' and 'Astern' directions is shown in Fig. 4.28 (b). At the end of travel of the cam follower, the link gets locked in position.

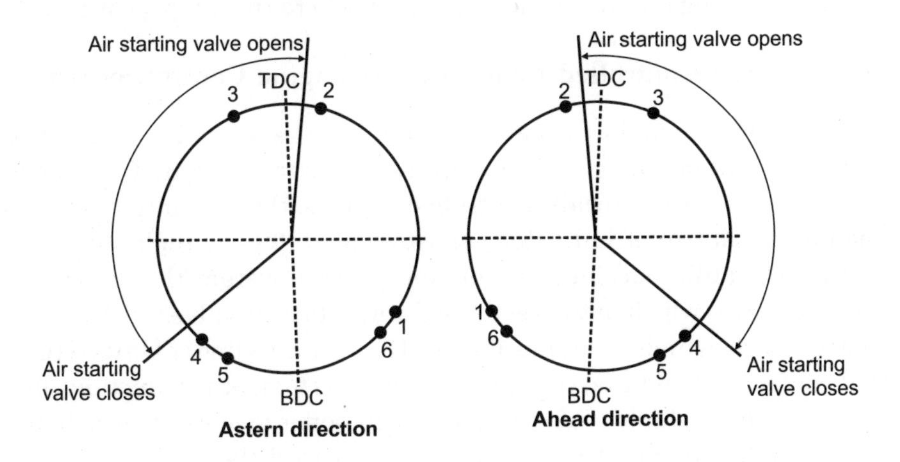

Fig. 4.28 (b) Timing diagram.

Various interlocks and blocking devices provide safe operation. As mentioned, the starting interlock prevents injection of fuel before starting and reversing sequences have been completed, the air/fuel interlock blocks starting air when fuel injection starts and the turning gear interlock blocks the starting sequence if the turning gear is engaged. Provided all the interlocks are released, starting air then reaches the appropriate cylinder to turn the engine in the desired direction. Once the engine picks up sufficient speed air supply is cut off and fuel is injected. Speed is controlled by speed signals sent to the governor according to the position of the speed-regulating lever in the control console. When the regulating handle is brought to the 'Stop' position, a shut-down signal is sent to the governor, which makes it move the fuel index to zero, thus activating the puncture valves on the fuel pumps and stopping the engine. Any engine shutdown fault will also activate the puncture valves. Additionally, an independent safety system slows down or stops the engine in case of a serious malfunction. Slow down is activated in the event of high temperatures in any of the engine cooling systems, high scavenge air or exhaust gas temperatures, low lube oil pressure and high crankcase oil mist. In the event that appropriate steps are not taken within stipulated time to correct any of these faults, a slow down air signal is sent to the governor which reduces the fuel pump index to slow down the engine. If the situation persists, a shut down air signal is then sent to the puncture valves on each fuel pump to stop fuel injection. Shut down is also activated in case of lube oil pressure dropping to a critical level, engine over speed or high temperature of the thrust bearing. Once the fault is corrected, the automatic shut down is reset by bringing the speed-regulating handle at the control console to the 'Stop' position.

Electronically Controlled Camshaft-less Engine Control System

Further developments in marine technology have made certain key engines parts of conventional engines quite redundant. In modern camshaft-less, electronically controlled engines, the conventional hydro pneumatic control of individual air-starting valves is replaced by an electro-hydraulic, mechatronic engine control system that activates solenoid valves which in turn actuate the air-starting valves with precise control. Each cylinder unit has its own Hydraulic Cylinder Units (HCU) that controls, in addition to the air starting system, other functions like fuel injection, cylinder lubrication and in some models even exhaust valve operation. Starting air distribution to different cylinders is controlled by individual solenoid valves that direct the pilot air to the air-starting valves – as opposed to the conventional cam-operated air distributor.

Fig. 4.29 shows the electro-hydraulic, mechatronic engine control system of M A N B&W ME-C engines that controls fuel injection, exhaust valve actuation, cylinder lubrication and other functions.

The main difference between the conventional MC-C engine and the new electronically controlled ME-C engines are the number of mechanical parts made redundant and replaced by hydraulic and mechatronic parts that significantly enhance engine functions. The parts replaced are, chain drive, camshaft with cams, fuel injection pumps, mechanical exhaust valves and actuators, starting air distributor, mechanical governor, mechanical cylinder lubricator etc. These have been replaced by the following:

Hydraulic Power Supply Units (HPSU)

Hydraulic Cylinder Units (HCU)

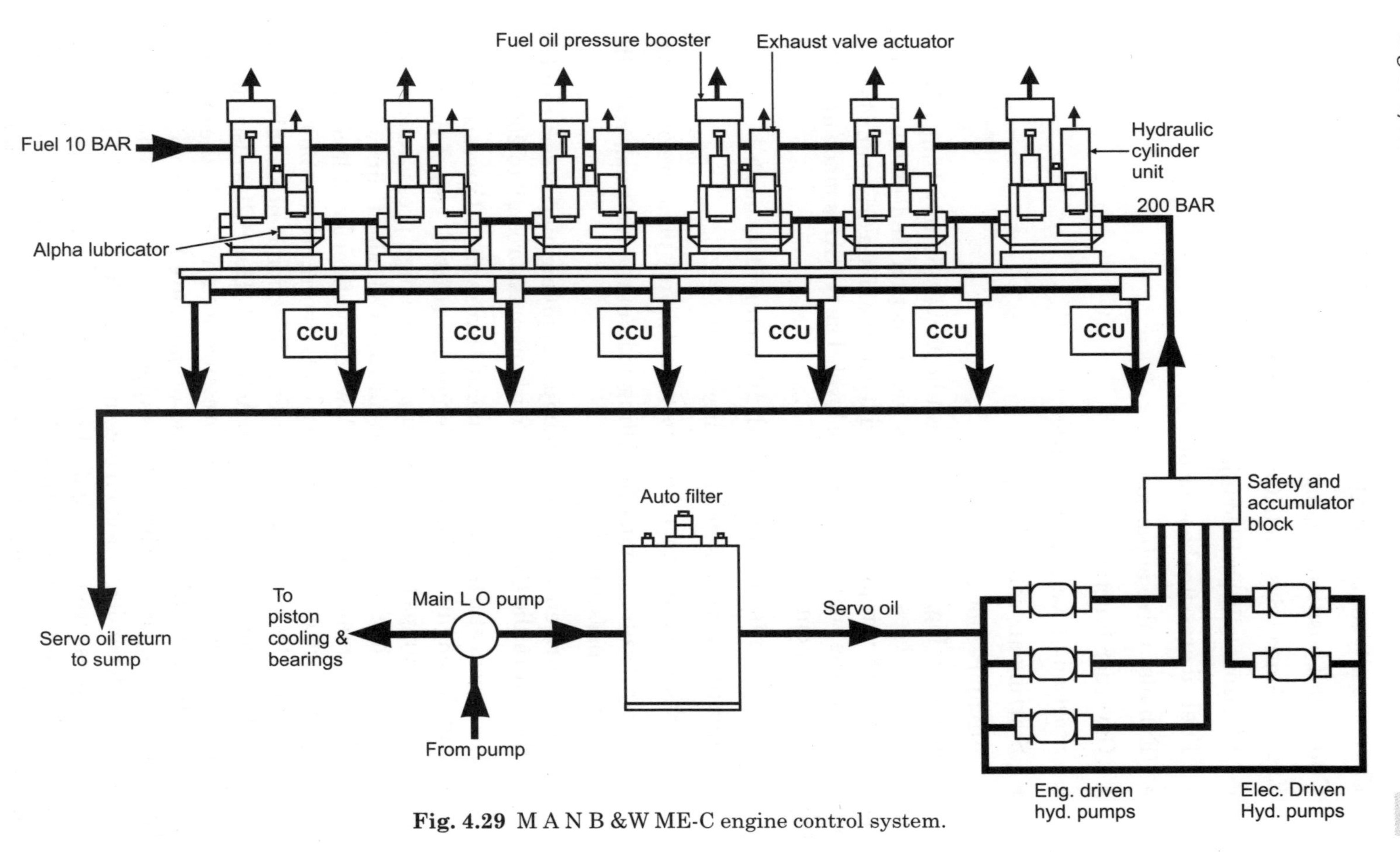

Fig. 4.29 M A N B &W ME-C engine control system.

Electronically controlled Alpha Lubricator

Crankshaft position sensing system

Engine Control System (ECS) that controls:

- Electronically profiled injection
- Fuel oil pressure booster pumps
- Exhaust valve actuation
- Governor functions
- Start and reversing sequences
- Starting air valves
- Auxiliary blowers

Referring to Fig. 4.29, the Hydraulic Power Supply Unit (HPSU) generates the necessary power for fuel injection and exhaust valve operation (which was previously provided by the chain drive) and located at the front of the engine at bedplate level. The HPSU is made up of the following components:

1. Auto-filter (self-cleaning) with 10-micron filter mesh
2. Line filters with 25-micron filter mesh
3. High-pressure start up pumps with delivery pressure of 175 bar
4. Low-pressure pumps for filling exhaust valve push-rod with delivery pressure of 4 bar
5. Engine driven axial piston pumps supplying high pressure oil to the Hydraulic Cylinder Unit with oil pressures up to 250 bar

The electrical start-up pumps generate hydraulic oil pressure prior to starting the engine and once the engine is started, two of the three engine-driven pumps take over the supply. The high pressure servo oil is then fed to each of the individual Hydraulic Cylinder Units which are mounted on a common base plate below the cylinder head level. These units, one for each cylinder, also carry two important electronic control valves namely, the ELFI (Electronic Fuel Injection control valve) and the ELVA (Electronic Exhaust Valve Actuator). The HCUs also carry a hydraulic oil distribution block with pressure accumulators, the exhaust valve actuator with ELVA and a fuel oil pressure booster with ELFI. This booster pump raises the fuel oil supply pressure from 10 bar to the specified load-dependent injection pressure of 600-1000 bar. ELFI controls the servo oil pressure to the fuel oil pressure booster, enabling precise control of the injection parameters (obtained in conventional engines as 'cam length', the 'cam inclination and angle') and even the number of activations per stroke. The ELVA for the exhaust valve actuator receives electronic actuator signals from the engine control system. The Alpha lubricator, also mounted on the hydraulic oil distribution block, is driven by the 200 bar servo oil pressure and provides precise control of cylinder oil feed rate, the cylinder oil being separated in the system from the servo oil.

Advantages of the Fully Integrated Control System

The advantages of the fully integrated control system of modern engines are quite comprehensive and are listed below:

- Better performance parameters due to variable electronically controlled timing of fuel injection and exhaust valves at any load.
- Optimum fuel injection pressure and rate at any load.
- Improved emission characteristics with smokeless operation.
- Better engine balance and equalized thermal load of cylinders.
- Lower rpm possible for manoeuvering.
- Lower steady running speeds due to sequential shut-off of fuel injectors, although all cylinders will be firing.
- Full power can be developed even with one fuel pump and one servo-oil pump inactive.
- Better acceleration, astern and crash stop performance.
- Integrated Alpha Cylinder Lubricators that allow greatly reduced oil feed rates.
- Control of exhaust valve timing allows optimum levels of scavenge air to be maintained by early closing of the exhaust valve when load decreases. This translates into low fuel consumption at low loads.
- Flexibility to run the engine at different modes of emission controls to comply with port requirements or to obtain minimum fuel consumption.
- Up-gradable to software development over the lifetime of the engine
- Better performance monitoring and engine diagnostics resulting in longer time between overhauls
- Longer intervals between overhauls. In some engines, overhaul intervals for components like liner, piston and piston rings can be as long as three years.
- Lower running and maintenance costs.

Unmanned Machinery Spaces (UMS)

Developments in automatic control systems and high reliability of equipments have resulted in centralising of control and supervisory functions. Modern ships employ a high degree of automation and the control room within the machinery spaces carries centralised instrumentation enabling monitoring of all control functions. Control consoles usually carry all important controls with instrumentation and operating buttons located for easy accessibility. Display panels have mimic diagrams with alarms lights located at relevant points, valve position indicators, pump running lights and indicators of all operating parameters enabling remote operation and monitoring of important functions from the centralised control room. Data recording systems are also provided to handle the large amount of information generated. For example, whenever a measured variable deviates beyond limits from the set value an alarm is sounded and the deviation and the time it occurred are recorded. Data logging can be set to run either automatically or at set intervals. The overall automatic control system employs a combination of electrical, electronic, hydraulic and pneumatic systems and devices interlinked with sensors, transmitters, controllers and regulators all monitored by computer systems. The result of this level of automation is unmanned or unattended machinery spaces. However, for UMS operation certain conditions must be satisfied. The essential requirements are:

1. *Bridge control*: To enable operation of the propulsion machinery from the bridge. Simple control and instrumentation are also provided to enable emergency engine control.

2. *Engine control room*: A centralised control room within the machinery spaces from where all main and auxiliary machinery can be operated and controlled. However, local manual control of essential machinery is mandatory.

3. *Alarms and fire protection*: An alarm system to provide warnings in the control room, engine room, the bridge and accommodation in the event of malfunctions and deviations of running parameters covering all important equipments and functions. A fire detection and alarm system covering the entire machinery spaces to be provided along with a fire extinguishing system that can be operated from a remote station outside the engine room

with facilities for control of emergency equipments and to operate emergency shutdowns.

4. *Emergency power:* A provision for emergency electrical power from a self-starting emergency generator to provide essential lighting. Such a generator is usually connected to a separate emergency bus bar, if not, automatic synchronising and load sharing is provided.

5. *Automatic bilge high-level alarm:* Sensors are provided in the bilge to sound high-level alarms and start the bilge pump automatically.

6. Adequate settling tank capacity.

Bridge Control

Since the various preparatory steps and sequence of events undertaken by engineers cannot be carried out from the bridge, bridge control must necessarily have built-in procedures and safeguards incorporated in the system. This would include not only correct timing and logical sequence of events but also protection and safety interlocks. The main engine could be controlled from both the bridge and the engine control room. However, the selection is made from the engine room so as to enable engineers to take control at any time, if required.

A programming and timing unit receives operating signals and carries out the correct sequence of events at the appropriate time. The timings are carefully controlled not only to avoid dangerous situations but also to initiate the next event in the sequence. Safety interlocks and other protection systems stop the progress of the sequence and also prevent engine starting in the event of abnormal conditions. Further, engine is shut down automatically in case a critical situation arises or a serious fault occurs. The various signals that initiate these events may be either electrical, pneumatic or hydraulic and appropriate transducers or pneumatic and hydraulic cylinders are employed to convert these signals into mechanical forces for actuation of various devices. For example, an electric governor receives speed signals from a tachogenerator and the corresponding adjustments to fuel control are made through a position transducer fitted to the fuel control rack.

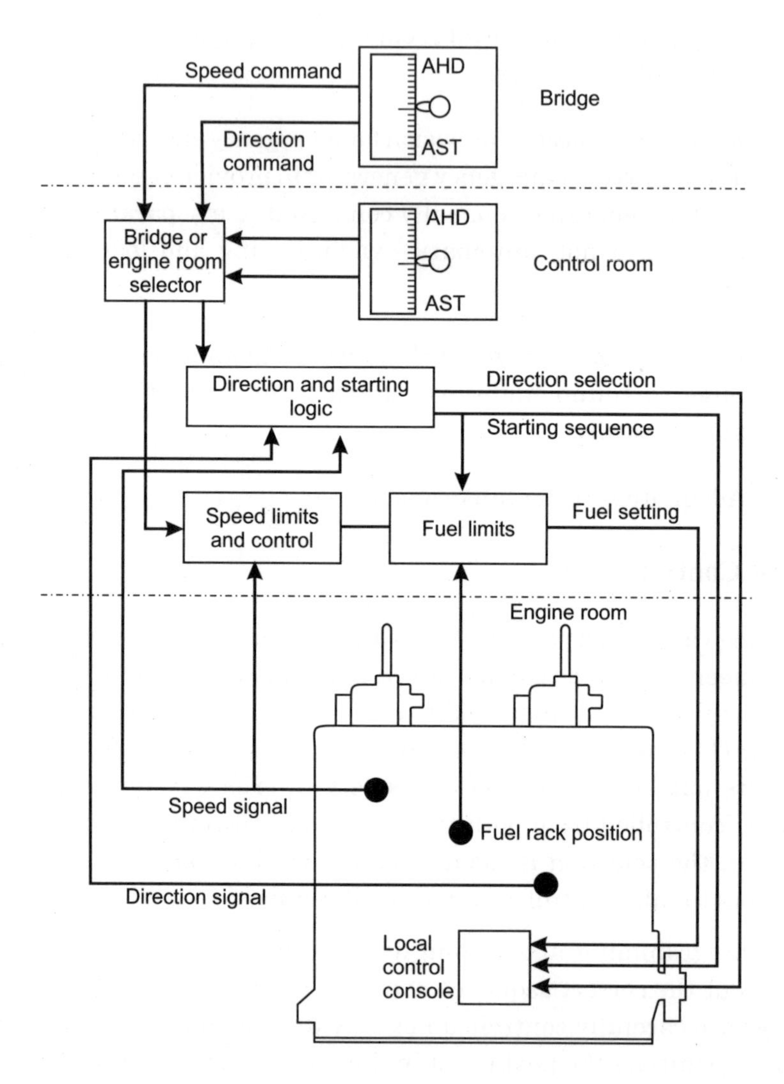

Fig. 4.30 Block diagram of a bridge control system.

A block diagram of a bridge control system is shown in Fig. 4.30. The bridge/engine room selector communicates operating commands regarding speed and direction of rotation, either from the bridge or from the control room, to the programming and timing units of the control system. The direction and starting logic control unit then sends signals to the camshaft positioner, to set the correct direction, and to the speed control unit, to send appropriate signals to the fuel pump rack positioner. Simultaneously, a logic device receives signals and triggers the starting sequence of the starting air system. The governor too receives a signal to release appropriate fuel for starting the engine and then to maintain supply to match the speed setting received from the control station. A feed back signal of the engine speed will allow the starting air to be shut off and enable the governor to maintain appropriate speed that would be displayed at both control stations.

The steering gear is bridge controlled that could be set for either automatic or manual operation.

Necessary checks and controls incorporated in the system to ensure safe operation of the engine are listed below.

1. To ensure that turning gear is disengaged.
2. To ensure that engine running direction is correct before releasing fuel.
3. To ensure that the engine is firing and picks up the required speed before shutting off starting air and to sound an alarm in case of start failure or failure to pick up required speed.
4. To avoid critical speed range and to set speed limits if abnormal conditions arise, for example in the event of a rise in cooling water temperature or deviations in other running parameters.
5. To monitor and maintain appropriate speed limits not only to ensure safety but also in the event of abnormal running conditions. For example, sudden acceleration is avoided in order to avoid excessive torque and speed is restricted to safe limits in the event of an excessive rise in cooling water temperatures or in order to match fuel supply to charge air supply from the turbochargers.
6. Immediate shut down of the engine in the event of a critical drop in lube oil pressure.
7. There is also a provision for automatic starting and load control of diesel generators. This would cover not only automatic synchronisation of the incoming generator with the bus bars and load sharing but also unloading and stopping of the stand by generator should the load reduce. The system might also incorporate controls for preferential tripping of non-essential loads when encountering emergency conditions, like the stand by generator failing to start.
8. Depending on the manufacturer there could be a number of other alarms fitted to signal various abnormal operating conditions.

❑❑❑

Chapter **5**

Hazards and Safety Devices

Certain hazards are inherent in a running marine engine despite all precautions and built-in safety features. Dangerous running conditions can develop in areas like the crankcase, scavenge spaces and starting air pipes. In addition, though rare, there is however the possibility of dangerously high pressures developing inside the cylinder liner or the engine running within its critical speed range. Built-in safety devices reduce the chances of a critical running condition developing and in case it does occur, they reduce the impact greatly.

Crankcase Explosion

For an explosion to occur, or indeed for a fire to begin, three elements are absolutely essential: a combustible material, air for combustion and a source of heat. Inside the crankcase of a running engine – or for that matter any fully enclosed, force-lubricated reciprocating machine including an air compressor – two out of the three elements are always present namely, potentially combustible oil spray and air. Therefore, it only needs the third element, a source of heat, to be introduced for a fire to begin and an explosion to occur. Fortunately, the oil spray – produced due to the constant agitation of the crankcase oil – is combustible only within a narrow band of temperatures with the potential to create a volatile concentration of oil and air. Oil spray normally present inside the crankcase is not combustible due to the large droplet size. Only if the temperature somewhere in the crankcase raises high enough to vaporise the oil, and if the vapour mixes with air in the correct ratio does the mixture become explosive. Two separate temperature regions have been found to create inflammable conditions: 270 – 350°C and above 400°C. This range has the potential to vaporise oil and produce mist particles of around 6 microns in size that, under certain conditions, ignite readily – in the presence of a heat source.

It is a **hot spot** that creates this heat source in the crankcase, and goes on to vaporise the oil in its immediate vicinity and also provide the spark to start an explosion. Hot spots arise due to overheating of any rubbing or sliding components like main or bottom-end bearings, stuffing box, timing chain rollers or sprocket wheels, gudgeon pins and piston rods. The reasons could be varied: incorrectly tightened cooling and lubricating oil pipes, too small bearing clearances, improperly assembled stuffing box, spurious spare parts or any abnormal interruption in the forced lubrication system. Crankcase explosions are known to have originated even at pistons (piston crown failure), piston rods, cylinder liners and even a fire outside the engine*. There is no recorded instance of a crankcase explosion occurring in the absence of a hot spot – the only exception being when spark or flame blew past the piston into the crankcase of a trunk engine. However, both two-stroke crosshead engines and four-stroke trunk engines are equally liable to crankcase explosions. And, size of the engine does not suggest proneness to explosion either, the risk being more or less same across all engine sizes.

Dynamics of Explosion

Oil vaporised by a hot spot circulates inside the crankcase to cooler regions and begins to condense, forming a mist. A mixture of this finely divided oil particles and air is combustible in certain concentrations. If this whitish mist is within the range of inflammability and now travels back to the hot spot, which, if it has reached a critically high temperature (about 500°C) with the continued generation of heat, the hot spot can then ignite the mist and a primary explosion can result. Thus, the hot spot which created the oil mist will become the ignition point for an oil

---

Recorded cases of crankcase explosion and causes (year-wise)

1995 Gearbox bearing
1996 Inlet pipe for piston cooling oil falling off due to incorrect tightening
1997 Piston rod interference with cylinder frame
1997 Incorrect spring mounted in piston rod stuffing box due to unauthorised spare part
1999 Fire outside the engine
1999 Weight on chain tightener falling off due to incorrect tightening
2000 Incorrect shaft in camshaft drive due to unauthorised spare part
2000 Main bearing
2000 Camshaft bearing
2001 Crankpin bearing
2001 Main bearing
2001 Crankshaft failure
2001 Piston crown failure
2002 Inlet pipe for piston cooling oil falling off due to incorrect tightening

mist explosion. The situation is compounded by inflammable gases such as hydrogen and acetylene released by the decomposition of oil during vaporising which have a lower flash point than the oil mist. Fuel dripping from a leaky injector finding its way into the crankcase of a trunk-piston engine can also create a low flash point mixture of fuel and oil – beyond a lube oil dilution of 10%, which is generally accepted as an allowable maximum. Even oxidation of lube oil after a long period of use can be a contributory factor.

The minor explosion will create a flame front and pressure wave that can travel fast across the length of the crankcase accelerating a combustion that can attain speeds of up to 1000 ft/sec with the possibility of a detonation*. The resultant shockwave has the power to rupture crankcase doors – unless relieved through a safety door. In case of a rupture before the explosion reaches its peak, unburnt mass of mist and air will be first expelled through the fractured door followed immediately by a burning mass. This will set up an immediate suction wave drawing air back into the crankcase due to the partial vacuum created inside (even if a primary explosion does not cause a ruptured door, air can still be sucked in through the crankcase vent in the absence of a non-return valve here). This fresh air will then mix with the vaporised and burning oil in the crankcase and set off a secondary or major explosion (pressure piling effect) causing widespread damage – often with dramatic and even sometimes fatal consequences.

There is also a secondary hazard. Oil mist expelled into the engine room by the pressure wave can ignite too if it meets a hot region like an improperly insulated exhaust pipe and start an engine room fire, causing widespread damage to electrical as well as mechanical equipments, not to speak of engine room personnel. Therefore, even if the pressure of a primary explosion is safely relieved through a crankcase door relief valve, and flame arrestors on the relief valve prevent a flame front from advancing beyond the crankcase, the danger of an engine room fire still exists.

Precaution therefore has to be the watchword to prevent crankcase explosions, the most crucial aspect of which is prevention of a hot spot. To achieve this, proper maintenance is crucial, which includes keeping lube oil in optimum condition, vigilance during overhauls and carefully

---

**The longer the combustion path, more violent the explosion. Large two-stroke engines with crankcase volumes in excess of 500 cubic metres are particularly vulnerable to serious damage due to this factor*

adhering to engine builder's instructions, use of only authorised spare parts, strictly following inspection and maintenance schedules, avoiding engine overload, regular checking of safety devices and alarms and general cleanliness. There is a possibility of lube oil dilution by fuel oil. However, this by itself is not found to be a causative factor for an explosion to occur and a consistent dilution of 8 to 10 per cent is unlikely to cause any serious harm, as earlier mentioned.

In the event a hot spot develops in a crankcase, watch-keeping personnel can detect it by the local rise in temperature, irregular engine noise or by the appearance of a white mist, apart from detection by temperature sensing probes installed in the crankcase. When overheating is suspected, sufficient time must be allowed for the heated crankcase to cool down before attempting to open a door. Otherwise, introduction of fresh air into the still hot crankcase can bring the oil-air mixture within the limits of inflammability.

Safeguards against crankcase explosion include: a suitably designed crankcase door of robust construction able to withstand a pressure of at least 12 bar, making sure that vent pipes are not too large and are led to a safe place on the deck, extending drain pipes from crankcase to reach well below the oil level in the drain tank and either subdivision of the crankcase or fitting a diaphragm mid length to prevent flame propagation.

There are several safety devices employed to detect a hazardous condition and to minimise or prevent damage if an explosion does occur. Oil mist detector, flame trap, crankcase relief valves and inert gas flooding system to inhibit flame propagation are some of these*.

Oil Mist Detector

This is a device that automatically detects oil mist in concentrations much below that required for an explosion. The apparatus samples air-vapour mixture drawn from the crankcase continuously and triggers an alarm if the concentration reaches the critical range allowing time to either slow down the engine or take other remedial measures.

**The devastating crankcase explosion on board the* Reina del Pacifico *on 11th September 1947 that killed 28 and injured many more led to the development of crankcase relief valves and flame arrestors. 143 crankcase explosions were reported between 1990 and 2001 by the Lloyds register of shipping alone, not counting minor explosions that go unreported. Of these, 21 explosions occurred in two-stroke engines and the rest in four-stroke, though this does not mean that four-stroke engines are more at risk.*

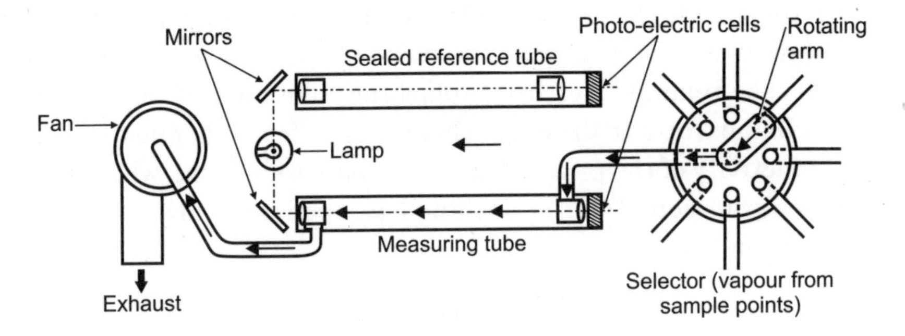

Fig. 5.1 Graviner crankcase mist detector.

As shown in Fig. 5.1 the device consists of a pair of photoelectric cells that measure mist density. Mist samples are drawn by a suction fan from each cylinder crankcase one at a time through a rotary valve and then a measuring tube. A light source at one end of the tube throws a beam across the mist in to a photo-cell fitted at the other end, generating a current proportional to the incident light, which corresponds to the mist density (oil mist at the lower end of the inflammable range has a high optical density). A separate reference cell with clean air sealed between lenses at both ends and with identical light source and photocell generates a reference current. Normally both the currents would be in balance, but if mist density increases, light falling on the measuring tube photocell will diminish and the resulting electrical imbalance will sound an alarm and show on a continuous chart record. A remote mist-density indicator if provided will show the density on a graduated scale with green, yellow and red sections. Further, the rotating sampler will stop, indicating the sampling point.

Crankcase Relief Valve

Fig. 5.2 shows an automatic self-closing crankcase relief valve that is generally fitted over a circular aperture cut into the crankcase door. These are provided on each cylinder crankcase on engines of moderate size and above while smaller engines have one at each end of the crankcase. These doors are essentially lightweight to reduce mass and inertia and afford a straight path for gas outflow without obstruction. An aluminium alloy lid held closed by a large diameter spring is attached to the crankcase door by a simple hinge. The rim of the lid carries an oil-resistant rubber ring to ensure a tight seal when closed. An aluminium deflector cover holds the valve spring in place as well as deflects the expelled gases safely toward the floor plates. The mouth of the valve is covered from inside the crankcase by a **flame arrestor** made of several layers of mild-steel wire gauze. The dome shaped cover projecting into the crankcase remains wet from splashing lube oil and makes for an excellent flame trap and dissipates heat quite effectively due to the high heat conductivity of the gauze. The large diameter spring, which makes the valve quite sensitive, is designed to open at a low pressure of 0.10 kgf/sq cm.

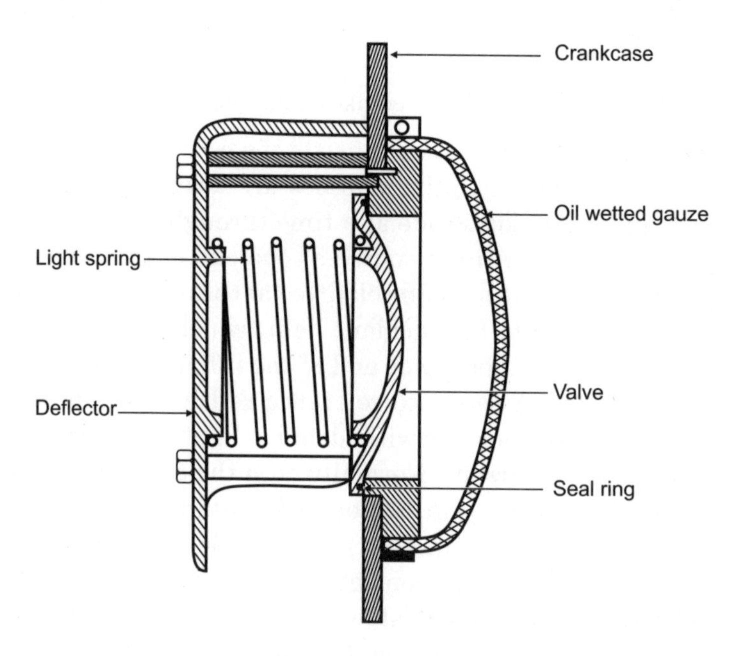

Fig. 5.2 Crankcase relief valve.

Inert Gas Flooding System

The scheme of continuous circulation of exhaust gas through the crankcase on its way to atmosphere – after scrubbing and cooling the gas – as a safeguard against potential crankcase explosion did not really take off due to the weight, cost and space requirements. However, an inert gas such as carbon dioxide with just 30% saturation by volume is found to be an effective safeguard against crankcase explosion. The gas ensures that oxygen content in the crankcase is reduced to a level (about 8%) where a flame front cannot propagate. In an inert gas flooding system, the gas is injected automatically into the crankcase when mist density enters the lower limit of inflammability. Or, the gas is manually injected when a critical situation is detected and an alarm is set off. However, on large diesel engines, use of inert gas creates a risk to personnel, as effective venting of the crankcase compartments is not easy and, entering an improperly vented crankcase is fraught with danger.*

Scavenge Fires

Defective piston rings or a badly worn liner will lead to piston blow past. This can be compounded by number of additional situations like defective injector, faulty fuel pump timing, lack of scavenge air, low compression, engine overload and chocked exhaust. Unburnt fuel and carbon then accumulate in the under piston scavenge spaces and the scavenge manifold. Additionally, if there is excessive cylinder lubrication, the excess oil can also find its way into scavenge spaces. The accumulated oil will carbonise with further heating and combined with the carbon and fuel already present forms an inflammable mixture that can be ignited by hot gases and spark from the piston blow past and cause a scavenge fire. Another source of heat that can trigger a fire is a blast of hot gases penetrating the scavenge spaces due to part blockage of exhaust ports on account of heavy carbonising and consequent reduction in area and rise in exhaust gas back-pressure (greater than scavenge air pressure).

The immediate result of a scavenge fire would be a drop in engine power, high exhaust temperature of the corresponding cylinder, surging of the effected turbocharger, smoke in the exhaust gas, smoke and spark from scavenge drains and irregular running of the engine. High temperature of the effected scavenge trunk will be another indicator.

If a fire is detected, engine must be slowed down; fuel shut off for the cylinder, cylinder lubrication increased and scavenge drains closed. A small fire will then burn itself out, but if the fire persists, the engine will have to be stopped and either carbon dioxide or dry powder fire extinguisher should be used through openings provided in the scavenge trunk. The trunking must not be opened immediately because of the risk of explosion. Turning gear should be engaged and the engine turned over to prevent seizure. After extinguishing the fire, trunking should be opened for inspection and the scavenge ports cleaned. At the next

**In on-going research by M A N B&W, an explosion suppression system based on the use of water mist is being studied. In addition to being safe, water can be removed in the course of the normal lubricating oil cleaning process. Pre-heated water under pressure injected through a specially designed nozzle, when a critical condition is detected in the crankcase, vaporizes and mixes with the surrounding relatively cooler gases and forms a mist. Though water mist cannot prevent an explosion, it reduces the impact considerably, thereby eliminating the risk of mechanical damage caused by the pressure wave.*

available opportunity, the liner, piston, rings, water seals, piston rod and stuffing box must be inspected for damages. The faults that caused the fire must be ascertained and rectified. Systematic and planned maintenance to ensure that piston blow past is prevented and that scavenge air pressure is always maintained higher than exhaust-gas back pressure and additionally, keeping the scavenge spaces clean and drained regularly is the best safeguard against scavenge fires.

Safety equipments include high temperature alarms fitted inside the trunking, set at about 200°C, and self-closing pressure relief valves fitted in the scavenge belt. Both must be periodically inspected and tested.

Starting Air System Explosion

As in the other explosion prone areas like crankcase and scavenge air spaces, there are two prerequisites for an explosion to occur in starting air pipelines: an inflammable mixture and a hot spot. Over years of running, lube oil from the cylinders and also from the air receivers accumulates in the starting air pipelines and the leakage of hot gases past a leaking or sticking air-starting valve provides the hot spot. The leaking hot gases can make the branch pipe connected to the starting air manifold red hot, vaporising the accumulated oil, and if the engine is stopped and restated before the pipe has time to cool down, the fresh charge of starting air mixes with the oil vapour and if the mixture is within the limits of flammability an explosion can occur. Starting air line explosions can be quite severe, blowing apart the entire starting air manifold.*

Another sequence of events that can lead to a starting air line explosion begins with a leaking fuel valve. Fuel oil accumulates in the cylinder and when the engine is then started the heat of compression pre-ignites the fuel and when the air starting valve opens, the cylinder pressure would be higher then the pressure of the incoming starting air leading to the burning combustion gases entering the starting air manifold and igniting the accumulated oil in the pipe line.

However, later investigations have revealed that auto-ignition of accumulated oil inside the starting air pipelines is the principal cause of a starting air line explosion, rather than cylinder backfires. The

** The infamous explosion on board the Cape town Castle in 1960 killed 7 people while the one in 1999 on board a container vessel during manoeuvring resulted in no causalities. However, Lloyds register database reports 11 incidents of explosions in starting air systems between 1987 and 1999 – all attributed to poor maintenance and unsatisfactory shipboard practices.*

incoming rush of starting air compresses the air already inside the branch pipe causing a temperature rise as high as 400°C which sometimes vaporises the entrained oil and causes self-ignition leading to an explosion.

To minimise the risk, once again, proper maintenance is the key. Starting air valves must be properly seated, ensuring air-tightness. Fuel valves must be maintained to ensure they are drip-free. And air receivers must be constantly drained during manoeuvring to minimise oil carry-over into the air pipe lines (for this reason, the compressor air intakes are usually located in an oil-free atmosphere).

Safety devices include flame arrestors fitted in the branch pipe immediately before the air starting air valve to restrict the risk of hot gases or burning fuel gaining ingress into the pipe and the manifold. A bursting disc with a safety cover is fitted to the starting air branch pipe leading to each cylinder in order to minimise damage in the event of an explosion. Where flame arrestors alone are used, a spring loaded relief valve is sometimes fitted to the manifold to relieve excess pressure.

Cylinder Relief Valve

This safety device is fitted to the cylinder cover to relieve excess gas pressure from the cylinder. Fig. 5.3 shows the half cross-section of a spring loaded cylinder relief valve. The valve made of stainless steel has a mitre seat on which the machined lower end of the spindle has a close seating. A shoulder at the top of the valve spindle restricts its lift. The lock nut at the top of the valve body allows setting the correct spring load on the spindle so that it lifts at not more than 20% of the designed cylinder pressure.

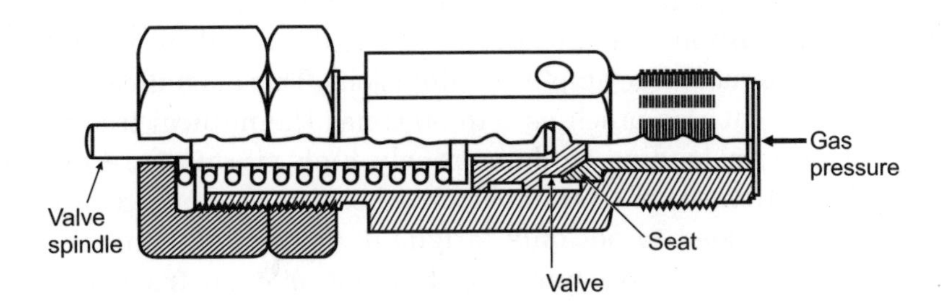

Fig. 5.3 Cylinder Relief Valve.

The valve seat area is designed to relieve only excess gas pressure and not intended to be a safety device in the event of starting an engine with undetected leakage of water or oil into the cylinder. As a safeguard, the engine is always turned over slowly using the turning gear to expel any leaked oil or water from the cylinder, as part of the engine stand by preparation.

If the valve lifts during manoeuvering, it could be due to excess fuel supplied at starting, a leaking fuel valve, a high fuel pump setting or improperly set fuel injection timing. In addition, if starting air is used for a 'crash stop' then high compression pressure near TDC may also cause a relief valve to lift. Valve lifting during running could be due to overloading the engine, timing faults, leaking or sticking starting air valves, or excess peak pressure due to incorrect conditions. While running, a hot air-inlet pipe will indicate a leaking starting air valve.

Critical Speed

Critical speed of a rotating body is the angular velocity which excites its natural frequency. This 'harmonic' excitation could result when pulsations in the engine torque coincide with the natural frequency of the shafting system which then begins to resonate. This critical resonance can lead to excessive systemic vibration. Once this critical stage is reached, the system is said to vibrate at its resonance frequency; resonance being the tendency of a system to oscillate at maximum amplitude when external pulsations acting on it reach its natural frequency. Excitation is generated by the engine due to its varying cyclic torques, on account of the varying forces and moments originating from the combustion pressure in individual cylinders and the inertia forces of the rotating and reciprocating masses. In addition, excitation can also be induced by the non-uniform reactive thrust from the water acting on the rotating propeller. Resonance speeds depend on the number of cylinders, the engine firing order and mass of the flywheel. A shaft system – made up of the shafting, rotating and reciprocating parts of the engine, bearings and propeller – may have a series of resonance speeds, the so called "barred speed range", at which it can produce large amplitude vibrations. This magnified frequency response can reach as much as 5 to 50 times the natural frequency of the shaft system leading to dangerously high vibrations. Control of torsional vibrations is of vital importance because excessive vibration of this nature can lead to not only fatigue damage to the engine or the connected hull structure but also damage or even fracture of the crankshaft, intermediate shafts or the propeller shaft. Continuous operation at the barred speed range is prohibited; allowing only a rapid passage through the range, which is specified for a given engine. Ships with electronic governors have a built-in critical speed control unit which operates on the speed setting signal, producing an automatic rapid passage through the barred speed range.

Fully-built crankshafts – with their heavy revolving masses – produce more torsional stresses than semi-built shafts. However, fully-built crankshafts with unequal crank angles overcome this problem. Reduction of the amplitude of torsional vibration is also obtained by increasing shaft stiffness and by fitting torsional vibration dampers to the crankshaft to dissipate the energy of vibrations. Additionally, axial vibration caused due to axial deflection of the shafting sets up varying axial forces on the shaft and is countered by installing axial vibration dampers at the free end of the crankshaft.

Controls and Alarms

Various controls and alarms are incorporated in the engine system to monitor the working of not only the different engine components but also pressures, temperatures of all engine operating systems and the multiple tanks holding cooling water and lube oils. Any deviation of working parameters beyond allowable limits triggers audible or visual alarms and if it is critical in nature, automatic cutouts shuts down the engine – offering an effective safeguard against seizure or overheating. In this event, cutout servomotors stop the engine by stopping the supply of fuel to the injectors.

Low-pressure alarms are incorporated in the piston cooling, jacket cooling and lube oil systems. The first two are generally set at about 2.5 bar while lube oil is set between 1.5 and 2.0 bar. In the event of failure of a circulating pump, the stand-by pump is automatically put into service and an alarm sounded. Automatic cutouts however operate at 2.2. bar (piston), 2.0 bar(cylinder) and 1.0 bar(lube oil).

High temperature alarms are also fitted in the piston, jacket and lube oil systems. Piston cooling water inlet alarm is set at 60°C, jacket cooling water cylinder outlet at 85°C and lube oil outlet at 50°C. Fuel oil low and high temperature alarms are generally set at 85°C and 120°C respectively in conventional engines. Additionally, low-level alarms are fitted to piston cooling water drain tank and jacket water expansion tank and low and high level alarms in lube oil sump tank.

Despite alarms, if corrective measures are not taken, engine shut down is triggered under the following conditions: lube oil low pressure of 1 bar, jacket cooling water high temperature of 96°C, thrust block high temperature of 85°C or camshaft low lube oil pressure of 1.5 bar. However, before these critical levels are reached, engine can be run at dead slow for some time. Nevertheless, emergency run commanded from the bridge can override engine shut down, by making the automatic cutouts inoperative and enabling the engine to continue running.

Other safety devices:

1. Flame arrestors are fitted to crankcase relief doors. Various inert gas or water mist is employed in the crankcase to combat fire. Crankcase relief valve is set at 0.05 bar, while the oil mist detector is set between 2 to 5 % LEL.
2. Temperature sensing probes in the scavenge air spaces give early warning of a fire. Fire extinguishing devices are also fitted. Scavenge space relief valves are set at 1.5 or 1.6 bar.
3. Temperature sensing probes fitted on bearings and thrust block indicate any abnormal rise in temperature
4. Temperature probes in starting air lines warn of abnormal high temperature in these areas. Starting air bottles carry relief valves set at 32 bar in addition to a fusible plug
5. Fire extinguishing systems are fitted in the exhaust manifold and trunking along with flame and spark arrestors.
6. Bursting disc, flame trap and relief valve in the starting-air line. Relief valve and low pressure alarm in control-air line.
7. Non-flow alarm and engine slow down in the cylinder lubrication system.
8. Low fuel oil temperature alarm.

❑❑❑

Chapter **6**

Engine Operation and Maintenance

Stand-by Preparations

Prior to starting the engine after overhauls or a long shut down, a thorough inspection of the engine must be carried out to ensure that no tools or cleaning materials are left in the engine. Blanks and locking devices wherever used must be removed.

The following check-list enumerates the preparations necessary to put a conventional engine on stand-by.

1. Check shut-off valves in the piston, jacket, injector cooling and the lube oil systems.
2. Check all tank levels and replenish where necessary.
3. Start another generator and synchronise with the running generator to take the additional load.
4. Start lube oil and all cooling water/oil pumps and maintain service pressures. Check for leakages and rectify.
5. Start up the main/aux boiler to raise steam.
6. Warm up the main engine to service temperatures.
7. Bleed air if any from the cylinder cooling water system, the turbocharger cooling system and air coolers by opening vents provided at all cooling water outlets. Trapped air can lead to not only fluctuating pressures but also generate stream, cause corrosion and local overheating.
8. Drain condensates from all scavenge air spaces and charge-air coolers.

9. Start preheating heavy oil settling and service tanks, start both oil purifiers and drain both fuel oil service tanks to remove sludge and water.
10. Prime fuel pumps and fuel injectors on all cylinders, while barring the engine over with the turning gear.
11. With the turning gear engaged, open indicator cocks and turn the engine over for few revolutions to check that all running components are free and also to expel any water, oil or fuel collected in the engine cylinders. Close all indicator cocks.
12. Ensure lube oil flows freely from all running gear bearings.
13. Check piston cooling water outlet funnels for normal flow, or the sight glasses provided at individual piston cooling oil outlets.
14. Ensure cylinder lubricating oil in the pipes between lubricators and quills are filled-up by manually turning the lubricators and checking the feed indicators. Continue turning for about 50 times while the engine is being turned over by the turning gear to ensure cylinder walls are adequately lubricated.
15. Check air bottle pressure and start air compressor to fill up the bottle if necessary. Drain air bottle and air pipes.
16. Inform bridge and try out engine room telegraph. Also check reversing servomotor by moving the telegraph from 'Ahead' to 'Astern' several times. Check running direction safety interlock in both directions. Ensure governor and cut-out servomotor respond satisfactorily to the speed-adjusting lever at the control stand.
17. Check the functioning of all automatic cut outs. With both the telegraph and running direction interlock set to the same running direction, set the supply pipe three-way cocks to 'drain', one by one, and ensure that load indicator moves to position 0, simulating an engine shut-down due to system pressure drop to a critical level.
18. Disengage the turning gear.
19. Open shut-off valve on the starting-air bottle. Set the automatic starting-air valve to the 'Automatic' position. Close vent cock on the starting-air distribution pipe.
20. Ensure all water, oil and fuel pressures are normal.

21. Inform bridge that engine is going to be tried out. Open all indicator cocks and blow through the engine, while watching the cocks. Shut off all indicator cocks.
22. Try out the engine in both 'Ahead' and 'Astern' directions.
23. Try out steering gear with both main and auxiliary steering systems. Check rudder position for correct response. Keep both steering gears ON.
24. Inform bridge that the engine is on stand-by.

Starting the Engine

1. For a Sulzer engine, set the speed-adjusting lever (fuel control) to a position sufficient for the engine to start firing and pick up speed (3.5 on the scale).
2. On receiving a starting order from the bridge, answer the telegraph by repeating the order on the engine telegraph. This will bring the reversing servomotor to the correct running position and make the hydraulic blocking device release the starting lever.
3. Pull the starting lever to the 'Start' position and release it as soon as the engine starts firing.
4. Adjust the required manoeuvring speed according to bridge order using the fuel control lever.
5. Check all system pressures and adjust where necessary. Keep air compressors in operation to maintain air bottle pressure during manoeuvring.
6. During this period, cooling water to charge air coolers must be reduced or shut off completely to avoid condensation inside the scavenge spaces.
7. On receiving a 'Stop' order, set the repeating engine room telegraph to 'Stop'. The fuel is automatically cut off to the engine by the cutout servomotor. Pull back the speed control lever once again to a position sufficient to start the engine. (As a rule the fuel control lever is brought to 0 once the ship, on an inward voyage, is berthed and manoeuvring is completed.)
8. For an 'Astern' order, answer the telegraph and repeat the procedure as before for starting the engine.
9. Start auxiliary blowers if necessary (in case it is not automatic) to supplement scavenge air during manoeuvring.

10. On an outward voyage, once manoeuvring is complete, raise engine speed gradually to service speed in accordance with bridge requirements.
11. Make an inspection tour round the engine; observe all temperatures, pressures and other parameters; look out for abnormal noises or vibrations and check the turbocharger speed.
12. On receiving 'Full Away', close the automatic starting-air shut-off valve, open vent cocks on the distribution pipe and drain the lines. Close the air bottle shut off valve. Switch off one steering gear.
13. On conventional engines, start injector cooling water pump and adjust temperature. Check heavy oil service tank temperature and bring to required level (90°C), drain the tank to clear sediments and sludge, ensure steam heating of fuel pipes is adequate and raise diesel oil temperature close to that of the heavy oil temperature. Put viscometer on auto control. Change over to heavy oil gradually, using the three-way change over valve. Maintain injector cooling water inlet temperature at 70°C and outlet at 90°C.
14. New generation diesel engines with common rail injection system are started on heavy oil and runs though manoeuvring on the same fuel. There is also no separate nozzle cooling system, since fuel oil itself fulfils that function.
15. Carry out periodic rounds of inspection as part of sea watch.

Watch Keeping

As a rule the relieving engineer reaches the engine room 10 to 15 minutes prior to taking over watch. An inspection round, starting from the funnel and extending to the bottom level, should include the following checks of the engine and related systems:

1. Step onto the funnel deck and observe the exhaust smoke. Normal smoke, indicating good combustion, would be light grey in colour. While a whitish smoke would indicate excess of air due to high charge air pressure or incorrect combustion; a blackish smoke, indicating poor combustion, could be due to any number of reasons: engine over load, excessive fuel delivery due to incorrectly adjusted fuel pump or injectors, leaking or eroded fuel injector nozzles, low scavenge air pressure, insufficient compression pressure due to worn or ill fitting piston rings or, defective scavenge air valves. Since the funnel

accommodates exhaust pipes from the main engine, the auxiliary engines and the boiler, it is a good practice to identify the source of the abnormal smoke. A sparking funnel on the other hand, would indicate a dirty exhaust gas economiser, engine running too long under low load or poor combustion. A fire in the economiser or a scavenge fire also cannot be ruled out.

2. Check jacket cooling water expansion tank level and if found low, replenish and monitor the level. Also, look for leakages in the system.
3. Check all settling and service tank levels, drain them, check heavy oil temperatures and ensure steam heating is adequate.
4. On the cylinder head platform, check jacket cooling water outlet temperatures and any leakages. Ensure air vent cocks on cooling water spaces in cylinder covers, turbochargers and air coolers are open. Check starting air pipes – high temperature could indicate a leaking air start valve. Feel fuel pipe connected to injector for the steady, rhythmic pulse. Check cylinder exhaust temperatures and any gas leakages from flanges etc. Check exhaust temperatures at inlet to turbine (500 ° C) and at exit.
5. Check scavenge drains for oil or water leaks.
6. Check scavenge air coolers – drains must show only condensate and not sea water. Air temperature should be low to ensure sufficient charge-air volume and efficient combustion. However, if it is too low (close to dew point) there is the risk of condensation and formation of rust inside the charge-air receiver. Maintain temperature above dew point by adjusting seawater outlet or the cooler by-pass valve. Check air pressure drop across the cooler. Too high a drop would indicate a fouled cooler.
7. Check hydraulic governor oil level.
8. Check fuel oil temperature and viscometer.
9. Check auxiliary engines – sump level, pressures and temperatures, leakages and abnormal noise.
10. Check feed rate of cylinder lubricators.
11. At the bottom level, check piston cooling water flow-indicator funnels or cooling oil sight glasses. Feel crankcase doors for abnormal rise in temperature or vibration to get a feel of the running gear, especially if any bearing has been replaced or re-scraped. Listen to the noise through the crankcase door for abnormalities like knocking.

12. Check all pumps for excessive gland leakage, vibration, abnormal noise and electrical load on the ammeter. Observe pressure gauge readings.
13. In the control room check all temperatures, pressures and other parameters. Check load on the diesel generators – avoid running on low load which can cause fouling. Any abnormal parameters noticed must be investigated and the problem rectified where immediately feasible. If it can wait, make a note for attending to it at the earliest possible. Temperatures must be brought to normal gradually to avoid thermal shock.
14. Check engine room logbook. Make a note of any problems in the previous watch, any cargo operations, adverse running conditions, any equipments isolated and special instructions.

In addition, the following checks on all auxiliary systems and equipments in the engine room should also form an essential part of the inspection.

1. Check the fuel oil and lube oil purifiers. Observe oil inlet temperatures, check for oil overflow, oil back pressure and the ammeter reading.
2. Check air compressor for normal operation, observe temperatures, check sump oil level, check air bottle pressure.
3. Check double bottom and sludge tanks, check bilge level, check stern-tube for oil level and pressure. Check thrust block and intermediate shaft bearings.
4. Check steering gear, observe oil level.
5. Check auxiliary boiler water level and blow through the gauge glass. Check boiler flame for correct combustion.
6. Observe any maintenance work being carried out on equipments or machinery.
7. Check working of the oily water separator and observe the condition of the water being pumped out.
8. Check the fresh water generator and observe the vacuum and fresh water output.
9. Take periodic rounds of inspection covering the main engine, auxiliary engines, boiler and all equipments and systems.
10. Be vigilant against potential fire hazards and critical or abnormal running conditions. Maintain log book, taking care to highlight any

abnormalities observed and action taken, while recording performance parameters.

11. Take note of all standing instructions and notices from the bridge, like oil transfers, ballasting / de-ballasting, bilge pumping etc.
12. Notify the bridge and inform the chief engineer promptly in the event of malfunctioning of any machinery or any emergency situation including fire, and then take corrective action.
13. Afford top priority to watch keeping and avoid undertaking any other work likely to interfere with it.

In accordance with the provisions of the current STWC convention, the officer in charge of the engine room watch (OICEW) must also be familiar with the following:

1. Communication system in the engine room
2. Escape routes
3. Safety and alarm systems
4. Fire fighting systems in the machinery spaces.
5. Arrangement and working of all machinery in the engine room.
6. Must be able to isolate, test and commission all machinery.

Additional vigilance and attention is needed when the engine is run under abnormal conditions. Following is a checklist:

Overload Operation

1. Keep a close watch on the load indicator position to ascertain degree of overloading and ensure it does not exceed the limit laid down in the engine trial reports.
2. Watch the exhaust gas temperatures after turbo-charger. It must not exceed the limit set out in the engine trial report.
3. Strong head winds and a fouled hull will also induce a heavy load and consequent speed drop. Increasing the speed will amount to overloading the engine. In any case, engine speed must not be increased beyond the maximum limit fixed on the speed adjusting lever at the control stand.
4. Cooling water and lube oil outlet temperatures must not exceed the maximum permissible limits.
5. Under overload conditions, cylinder lubricator throughputs must be increased using the delivery rate adjusting lever.

6. Inspection rounds must be more frequent and all temperatures, pressures and other running parameters must be observed more frequently.

Operating at Minimum Speed

1. Cylinder lubricator throughput must be reduced or brought to the minimum if load is very low.
2. Cooling water and lube oil temperatures must be maintained at normal levels.
3. Cooling water to charge air coolers must be reduced or shut off completely to avoid condensation inside the scavenge air spaces.
4. At very low speeds (about $1/6^{th}$ of rated speed, just enough to keep the cylinders firing), governor will tend to become insensitive to speed. Engine speed then should be maintained by fine control of the speed-adjusting lever at the control stand.

Running Problems

Smoky Exhaust: This is a result of poor combustion and could be due to one or more of the following:

1. Engine overloaded. Check load indicator and exhaust gas temperatures.
2. Too low charge-air pressure due to fouled air suction filter or blower. It could also be due to fouled charge-air cooler, fouled reed valves, partly blocked scavenge ports, or fouled exhaust ports. Low turbocharger rpm due to leakage of exhaust gases before turbine, high back pressure after turbine, or a fouled turbine could also be reasons. In this event exhaust temperatures would rise.
3. Excessive fuel-oil delivery to one or more cylinders.
4. Insufficient compression pressure due to worn liner or piston rings.
5. Unsuitable fuel or low fuel temperature.
6. Scavenge fire, indicated by local overheating and high exhaust temperatures.
7. Incorrect atomisation due to worn, eroded or partly blocked injector nozzle.
8. Excessive cylinder lubrication.

Piston Knock at TDC

1. The affected cylinder is overloaded. Check fuel pump delivery stroke.
2. Early fuel injection. Check corresponding fuel pump timing and cam.
3. Sticking fuel injector needle.
4. Badly fouled cylinder.
5. The top piston ring butts against the ridge worn on the top of the liner wall.
6. Excessive clearance between liner and piston.
7. Loose running gear bolts of the affected unit.
8. Connecting screws of the piston rod or the piston rod nut are not tight.
9. No clearance between piston crown and cylinder head.

Fluctuating Cylinder Cooling Water Pressure

1. Air accumulation in the cooling system due to inadequate venting. Ensure all vent cocks are open.
2. Too low water level in the expansion tank or an empty tank, leading to insufficient static pressure on the system. Replenish the tank and vent the system.
3. Clogged cooling water pipes and partly closed regulating valves.
4. Ingress of gas into the cooling water system from a cracked cylinder head or liner.
5. Air ingress into the cooling water-circulating pump through poor gland packing.
6. Insufficient water inlet to pump due to throttled valve in the return pipe.

Most of the above events could also lead to high cooling water temperature.

High Cylinder Cooling Water Temperature

1. Engine overload or piston running hot.
2. Valves in the cooling water lines may be either shut or insufficiently vented.

3. Crack in cylinder head, liner or exhaust valve cage.
4. Temperature controller may be malfunctioning.
5. Piston is running hot and may be seizing

Piston Running Hot

Indications include cylinder knocking at both ends of the piston stroke, rise in both piston and cylinder cooling water outlet temperatures of the affected unit and dropping engine speed.

Remedial measures:

1. If the temperature rise is not very high, check thermostatic valves if present, or check heat exchanger inlet/outlet temperatures, check cooling water/oil pressure.
2. Check fuel injection for after-burning.
3. Ascertain if there is piston-blow past due to excessive ring clearance or worn liners.
4. If problem persists and the temperature continues to rise, cut-out fuel to the affected cylinder. Increase cylinder lubrication for that unit.
5. Stop the engine if possible and allow the affected unit to cool down. Continue cylinder lubrication manually and turn over the engine to prevent seizing.
6. Take precautions to prevent fire in the under-piston scavenge spaces.
7. Open the unit and dismantle piston. If scoring is not heavy, smoothen surface with oilstone, if not replace piston. Re-check cylinder lubrication throughput.

Running Gear Bearing Temperature High

1. Low lube oil pressure due to number of possible reasons like low sump level, trapped air in the lube oil system, defective lube oil pump, clogged filter, etc.
2. Lube oil cooler is clogged or seawater flow is inadequate.
3. Bearing clearances are either too low due to inappropriate bedding or too high due to excessive bearing wear.
4. Lubricating passages or grooves are clogged.
5. Oil leaking through a loose joint.

6. Partly closed shut off valve or faulty regulating valve.
7. Presence of water or other impurities/hard particles in lube oil
8. Inappropriate tightening of the bearing covers resulting in deformation.
9. Crank-web bearing pin or main bearing journal has suffered damage.

Remedial measures:

1. Increase lube oil pressure to the extent possible; but if problem persists, stop engine and allow to cool down.
2. Take precautions against crankcase fire.
3. Open up the affected bearing.
4. If the damage is slight, repair/recondition the journal and bearing, if not replace as required.
5. If repairs cannot be carried out immediately, shut off fuel supply to the affected unit and continue voyage at reduced speed.

Lifting Cylinder Relief Valve

1. High fuel pump setting.
2. Sudden overload in rough weather.
3. Pre-ignition.
4. Burning of accumulated fuel in the cylinder due to a leaking injector.
5. Leaking or sticking air start valve.
6. Water or oil accumulation on the piston crown.
7. Incorrect setting of relief valve lifting pressure.

High Exhaust Gas Temperature

1. Insufficient charge air due to fouled turbocharger or air cooler, fouled air filter or scavenge reed valves or partly blocked scavenge or exhaust ports.
2. Unsuitable fuel or incorrect fuel injection timing.
3. Thermal overload due to a scavenge fire.
4. Piston blow past in one or more cylinders.
5. Leaking exhaust valves.

Rising Lube Oil Sump Level

1. Rolling and pitching or listing to one side.
2. Piston or jacket cooling water leaking into the crankcase.
3. Valve from the lube oil storage tank inadvertently left open.
4. Wrong position of lube oil purifier valves. Check inlet /outlet valves.

Drop in Engine Speed

1. Defective fuel pump or fuel injector leading to variation in fuel injection.
2. Incorrect fuel injection or bad quality of fuel. (Check priming plug on injector).
3. Fouled scavenge air or exhaust passages or ports. Defective scavenge air valves.
4. Scavenge fire.
5. Governor malfunctioning or worn out linkages. Check governor setting or bleed air if present.
6. Air lock in the fuel lines or presence of water in fuel.

Engine Over Speed

1. Faulty governor.
2. Sticking fuel pump racks.
3. Propeller emerging from water in rough weather leading to racing. Normally, governor maintains equilibrium to minimise speed fluctuation.

Irregular Running of Engine

1. Defective fuel circulating pump.
2. Irregular firing of one or more cylinders.
3. Air trapped in fuel lines.
4. Fuel temperature either too high or too low.
5. Sticking or defective governor.
6. Fluctuation in charge air pressure due to defective, fouled or surging turbocharger.

Port Stay and Anchorage

During port stay, either at anchorage or alongside berth, if the vessel has to be in readiness to move on its own power at short notice then following procedures are followed:

1. Main engine jacket cooling water circulation is kept on and linked to the generator cooling system, to keep the main engine warm.
2. No maintenance work is undertaken that will interfere with manoeuvring.
3. Sea watches are maintained.
4. All regulations for prevention of pollution and for ensuring ship's safety are observed.

If the vessel is alongside berth for a longer stay or at safe anchorage, these steps need to be taken:

1. Shut down main engine and the auxiliaries not required.
2. Depending on the electrical load shut down one generator.
3. On a tanker get the cargo pumps ready for operation.
4. Follow correct procedures and safety precautions while receiving on board fuel and lube oil and while transferring oils.
5. Turn over the main engine by turning gear occasionally as feasible during port stay, while running the lube oil-circulating pump.
6. Carry our planned maintenance routines.
7. Attend to surveys and inspections by regulatory bodies.

Safety Measures to be Taken Prior to Engine Overhauls

1. Ensure that both the automatic starting-air shut off valve and the shut-off valve on the air bottle are closed.
2. Vent all starting air pipes and keep the vents open.
3. Open indicator cocks on all cylinders and leave them open during overhaul.
4. Keep the turning gear engaged and secure the hand lever.
5. Before opening crankcase doors follow all safety procedures.

Maintenance

Maintenance Schedule for Conventional Engines

Builder's manuals provide specific guidelines for engine maintenance. A maintenance schedule based on running hours for Sulzer RND engines is given below:

1. At 150 running hours
   - Oil / grease regulating linkages.
   - Carry out routine maintenance of turbochargers according to manual.
   - Clean charge-air coolers or earlier if pressure-drop exceeds the limit.
2. At 1500 running hours
   - Check piston cooling water telescopic pipe glands.
   - Drain oil from Woodward governor, rinse with cleaning oil and replenish.
   - Carry out routine maintenance according to turbocharger manual.
   - Check gear wheels and lube oil spray nozzles of camshaft.
   - Check fuel injectors. Clean and readjust injection pressure.
   - Check air-starting valves and retighten valve spindle.
   - Check main bearing and retighten bolts.
   - Check all crankcase connections and retighten where necessary.
3. At 3000 running hours
   - Check the pre-stressing of tie rods.
   - Check clearance of main bearings and adjust where necessary.
   - Check axial play of thrust bearing.
   - Check cylinder liners and clean scavenge and exhaust ports.
   - Clean under-piston scavenge spaces.
   - Check fuel injector nozzles for wear.
   - Check crosshead bearings and bottom-end bearings. Adjust clearances.
   - Check crank web deflection.
   - Check piston cooling water telescopic pipe glands ands replace rings.

4. At 6000–8000 running hours
    - Inspect cylinder liner, gauge the bore and remove ridge from the top.
    - Check cylinder lubricating quills for function and water tightness.
    - Check copper sealing rings of cylinder heads.
    - Check clearance of all crankshaft bearings.
    - Inspect crankcase, clean and check tightness of all screwed connections.
    - Check piston rod stuffing box. Check ring clearances and replace as required.

Although engine maker's recommendation of maintenance based on running hours is a valuable guide, frequency of overhauls can also be decided based on calendar-time that is applicable across all makes of engines. Overall, this affords a better control of the maintenance function to the ship owners, who are better placed to plan a maintenance strategy, leaving the tactical details to the ship's personnel.

It is to be noted that maintenance tasks on a particular piece of machinery would generally include inspection, minor overhaul, major overhaul and periodic surveys. A maintenance plan based on calendar time is give below.

At one-and-half months interval

- Check fuel injectors. Clean nozzle and adjust injection pressure.

At six months interval

- Check and clean exhaust manifold.
- Inspect the piston. Check the condition of rings and inner cooling spaces.
- Check and clean cylinder liner scavenge and exhaust ports. Gauge the bore to measure wear. Check cooling water spaces and clean by flushing with water. If scales cannot be removed fully, withdraw liner for cleaning.
- Examine piston-cooling system.
- Inspect heat exchangers of the piston and jacket cooling water systems.
- Check clearances of crosshead and bottom-end bearings.
- Overhaul and test starting-air valves.

At one-year interval

- Check clearances of crank shaft main bearings. Check shaft alignment. Clean lube oil channels on bearing inserts.
- C heck clearances of thrust block.
- Clean starting air pipelines and air bottles. Stream oil to remove deposited oil.
- Examine lube oil system and clean as recommended.
- Examine engine manoeuvring gear for play or linkages. Oil/ grease as required.
- Examine governor linkages for play. Clean and oil connections.
- Check fuel pumps and adjust if necessary.

Planned Maintenance System

The maintenance schedules enumerated above are based on subjective analysis of normal breakdowns of ship's equipments, and devised to ensure that breakdowns are kept low, at an acceptable level. However, it is imperative that a balance be found between breakdowns and scheduled maintenance, because while breakdowns and repairs cause machine downtime and consequent revenue loss, scheduled preventive maintenance also costs money in terms of labour, spare parts, consumables, overheads and other direct and indirect costs. A planned maintenance system strives to achieve this balance, ensuring that the best possible maintenance policy is in place, in terms of net cost to the company, by minimising the total operating cost. Employing modern management techniques like work study, organisation and methods, budgetary control, operational research and detailed scientific analysis of extensive data collected on spare gear consumption, man hours required for specific tasks and patterns of equipment break downs, have produced an optimal and integrated planned maintenance system that is increasingly being adopted on modern vessels. The introduction of a planned maintenance (PM) program would first and foremost bring a scientifically organised work schedule. This would result in number of benefits: increased effectiveness of the ship's workforce (with consequent reduction in the work carried out in dry dock), reduction in spare parts consumption, reduction in the overall time required for a specific task, high performance levels of ship's equipments, reduction in loss of hire of the vessel due to breakdowns and higher safety standards. PMS also incorporates machinery surveys and frequency of overhauls as required by classification societies. For example where an approved PMS is in use, some classification societies might negate the need to survey an item that has fallen due based on a review of its condition as per the PMS records. Alternately, if a piece of equipment is due for an overhaul according to the PMS, then it may be surveyed at the same time even if it is not yet due.

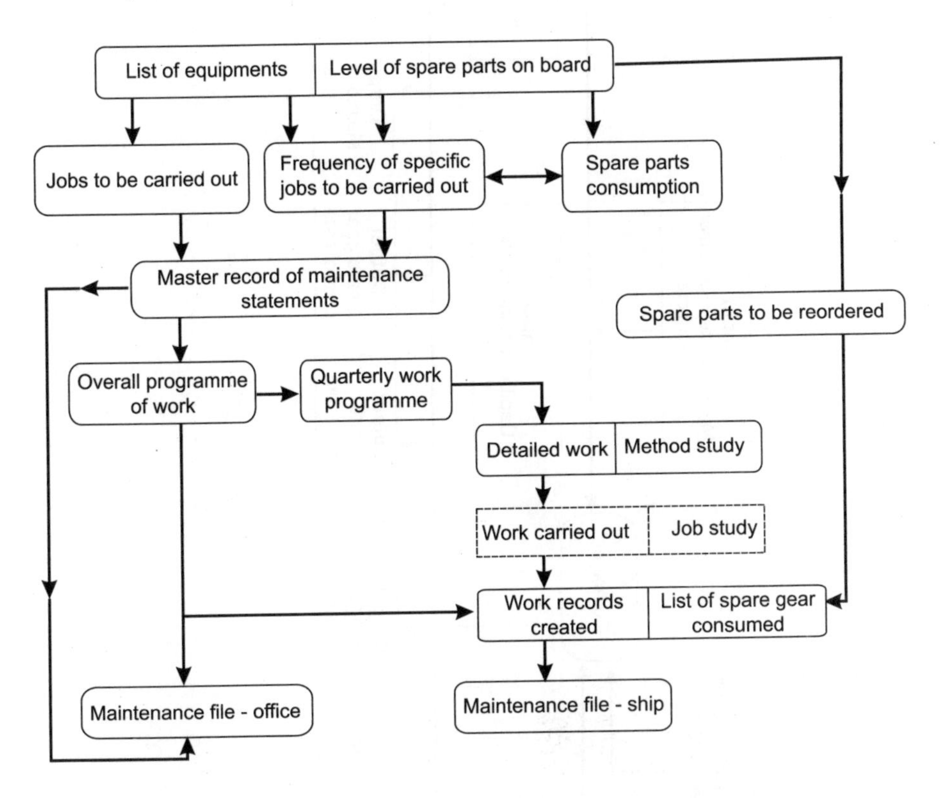

Fig. 6.1 Flow chart of a Planned Maintenance System.

Fig. 6.1 shows the flow chart of a computerised planned maintenance system. As can be seen, the main data consists of the list of equipments to be maintained and the level of spare gear inventory. The program software is based on detailed techno-economic analysis and commercial evaluation and defines in detail the work to be carried out on each piece of equipment, including the sequence of each individual task, frequency of inspections and overhauls and a list of spare gear consumed. A record of these events is maintained in order to keep the system updated with respect to the overall work programme. This facilitates periodic review of the program by the ship owners for necessary correction or modifications. The system thus works as an integrated maintenance control system that also enables the ship owners to prepare a budget for ship's maintenance and overall operation.

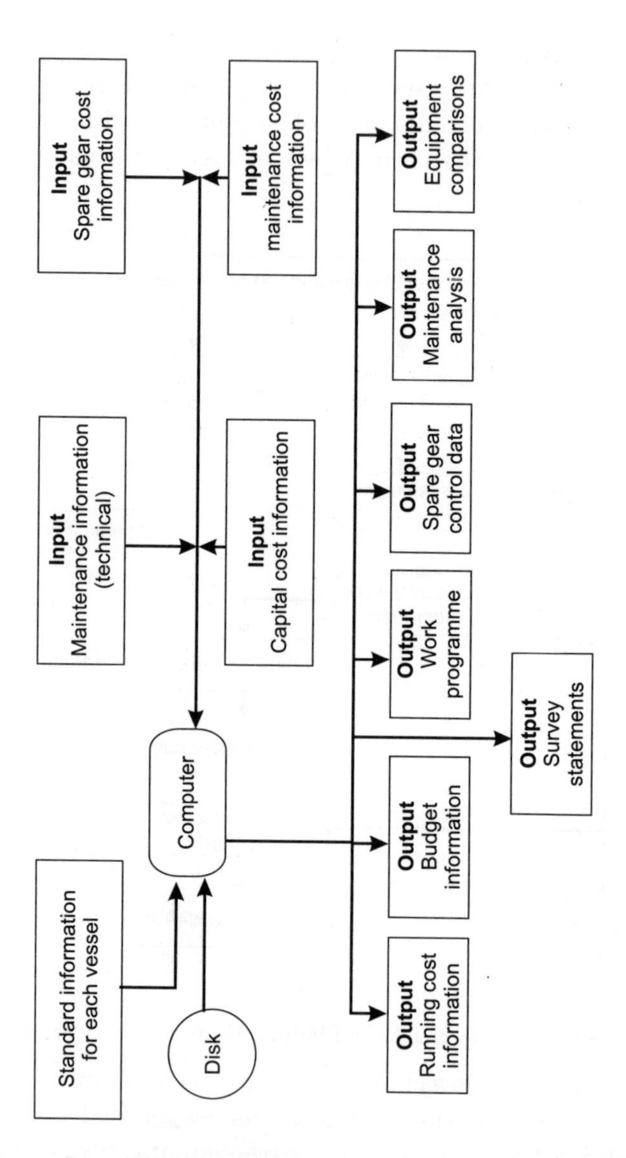

Fig. 6.2 General outline of a computerised maintenance system.

Fig. 6.2 gives a general outline of a computerised system. The system stores various kinds and levels of information concerning the ship and its machinery and provides output that helps in the efficient management of the maintenance and operation of the ship. The input data includes both master details of the vessel that contain standard information and itemised technical details of sub-assemblies and individual parts. The system in turn returns explicit information on diverse areas like work programmes and schedules, spare part control including re-ordering, status of equipments considering the ship's age, operating cost, reports of maintenance analysis and survey statements.

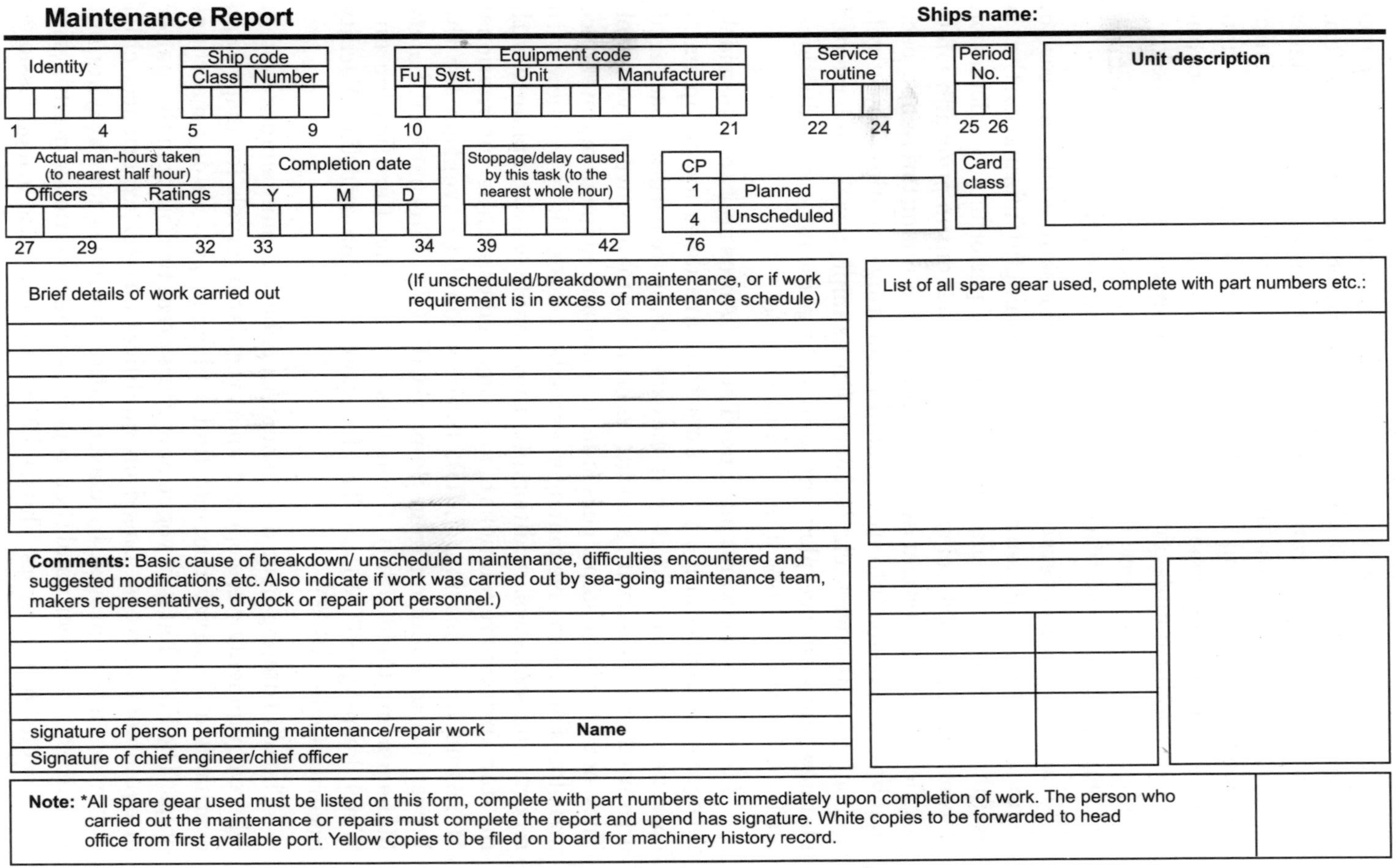

Maintenance Report **Ships name:**

Identity (1–4) | Ship code: Class, Number (5–9) | Equipment code: Fu, Syst., Unit, Manufacturer (10–21) | Service routine (22–24) | Period No. (25 26) | **Unit description**

Actual man-hours taken (to nearest half hour): Officers (27–29), Ratings (–32) | Completion date: Y, M, D (33–34) | Stoppage/delay caused by this task (to the nearest whole hour) (39–42) | CP: 1 Planned, 4 Unscheduled (76) | Card class

Brief details of work carried out (If unscheduled/breakdown maintenance, or if work requirement is in excess of maintenance schedule)

List of all spare gear used, complete with part numbers etc.:

Comments: Basic cause of breakdown/ unscheduled maintenance, difficulties encountered and suggested modifications etc. Also indicate if work was carried out by sea-going maintenance team, makers representatives, drydock or repair port personnel.)

signature of person performing maintenance/repair work **Name**

Signature of chief engineer/chief officer

Note: *All spare gear used must be listed on this form, complete with part numbers etc immediately upon completion of work. The person who carried out the maintenance or repairs must complete the report and upend has signature. White copies to be forwarded to head office from first available port. Yellow copies to be filed on board for machinery history record.

Fig. 6.3 Sample of maintenance report form.

To ensure systematic data collection with regard to the maintenance work, information is collected through standardised forms. A sample form is shown in Fig. 6.3. As can be seen, the form is scientifically designed and organised and uses indexing based on classification of ships equipments and systems into different categories, to facilitate data collection in an ordered manner. The form collects both master data and itemised specific details on the following: equipments data, breakdown details, man-hours, work details, spares consumed etc. This ensures systematic storage and retrieval of data for future reference.

The data collection system therefore produces an accurate record of engine operating condition and provides meaningful information that is useful in various ways: planning, analysis of performance, calculation of wear rates, power developed, fuel and lube oil consumption and safety certifications. To summarise, PMS provides a quantitative approach enabling management to study a ship's operating conditions objectively and on which to base their decisions. For example a well-implemented planned maintenance plan can provide significant cost reductions: a saving between 4-10% of maintenance costs.

Maintenance softwares like **SpecTec** and **Amos** are two of the widely recognised maintenance programs that are designed to optimize operational efficiency. To cite just one example, spare parts are packaged, organized and labeled with accurate descriptions to ease inventory control. They not only help ship's personnel to plan and manage maintenance and stores, but also give management ashore a complete overview with reports and statistics. Additionally, they enable easy compliance with the standards and regulations of maritime bodies such as, International Maritime Organisation (IMO), International Convention for the Prevention of Pollution from Ships (MARPOL), International Safety Management Code (ISM), Standards of Training, Certification and Watch keeping (STCW) and The International Ship and Port Facility Security Codes (ISPS).

The old running-hour or calendar-time based maintenance system where equipments were opened up at regular intervals for inspection, overhauls and survey does not find a place in present-day merchant ships. The PMS that is in vogue now offers great flexibility in the matter of equipment surveys. Based on qualification and experience, chief engineers are approved by classification societies for carrying out machinery inspections and their reports on the condition of machineries are accepted by surveyors. Thus, if PMS is implemented in a professional and optimal manner, time between inspection, overhauls and survey can be extended greatly. Secondly, technological advancements have brought in new materials, methods of manufacture and more efficient and innovative designs of various engine components that offer much enhanced service life. Notwithstanding all this, managements tend to have their own philosophy on maintenance: breakdown maintenance *Vs* preventative maintenance *Vs* condition based. Which system is adopted and how well it is implemented will depend on this philosophy.

Following is a general time chart of overhaul intervals, along with expected service life of various engine components for a modern engine maintained expertly according to a planned maintenance system.

Engine component	Overhauling intervals (hrs)	Expected service life (hrs)	Remarks
Piston crown	12 – 16,000	40,000 – 80,000	Pressure test at every 2nd overhaul.
Piston rings	12 – 16,000	40,000 – 80,000	Varies according to engine model.
Stuffing box	12 – 16,000	12 – 24,000	Renew lamellas as required.
Cylinder liner	12 – 16,000	40 – 80,000	Liner condition to be checked once a month through scavenge ports.
Exhaust valve	3 – 6000 Inspect and grind seats as required	60 – 100,000	Spindle may be reconditioned by rechroming.
Actuator gear	32,000 (Hydraulic system)	64,000	
Fuel injector	8000, depending on fuel quality	8000 (nozzle) 16,000 (guide)	Overhaul and replace nozzle every 8000 hrs.
Fuel pump	16,000	40,000 Renew	Change sealing rings or recondition on barrel, plunger and valve.
Cylinder cover	96,000		Check for burning at fuel injector holes.
Air-starting valve and relief valve	12,000	Same as engine life	
Cylinder lubricator (Mechanical)	16,000	Same as engine life	Clean oil tank, filter and pump. Check timing.

Cont....

Engine component	Overhauling intervals (hrs)	Expected service life (hrs)	Remarks
Alpha lubricator	16,000	Same as engine life	Refill accumulator
All bearings	6 – 8,000 Check clearances and c/s haft deflection	64,000 X-head brg. 96,000 All others	Inspect as per classification requirement
Chains	3 – 4,000 Retighten	96,000	
Chain wheels	3 – 4,000 Inspect	Same as engine life	Inspect and retighten every 500 hrs.
Reversing and regulating gear	3 – 4,000 inspect moving parts	Same as engine life	Change oil every 4000 hrs.
Tie rods	6 – 8,000 retighten every year	Same as engine life	
Holding down bolts	6 – 8,000 retighten every year	Same as engine life	
Turbocharger	Dry clean once a day	Follow makers recommendation	Check makers recommendation
Air coolers and air filters	Clean as required	40 50,000	Clean when diff. pressure increases by 50 % over sea trial value
Reed & butterfly valves in scavenge air manifold	Check during scavenge port inspection	Same as engine life	
All oil filters	Clean as required		
Lube oil sump tank	Clean every 16,000 hours		

❑❑❑

Bibliography

1. Pounder, C.C. *'Marine Diesel Engines'*, 1972.
2. Thomas D. Morton and Leslie Jackson. *Reed's Motor Engineering Knowledge for Marine Engineers,* Volume 12, 2002.
3. Wharton, A.J., *Diesel Engines,* 2001
4. Tambwekar A.S., *Watch keeping for Marine Engineers,* 2003

Index

H

I

J

L

M

O

P

R

S

T

U

V

W